Science and Fiction

Editorial Board

Mark Alpert
Philip Ball
Gregory Benford
Michael Brotherton
Victor Callaghan
Amnon H Eden
Nick Kanas
Geoffrey Landis
Rudi Rucker
Dirk Schulze-Makuch
Rudy Vaas
Ulrich Walter
Stephen Webb

For further volumes:
http://www.springer.com/series/11657

Science and Fiction – A Springer Series

This collection of entertaining and thought-provoking books will appeal equally to science buffs, scientists and science-fiction fans. It was born out of the recognition that scientific discovery and the creation of plausible fictional scenarios are often two sides of the same coin. Each relies on an understanding of the way the world works, coupled with the imaginative ability to invent new or alternative explanations—and even other worlds. Authored by practicing scientists as well as writers of hard science fiction, these books explore and exploit the borderlands between accepted science and its fictional counterpart. Uncovering mutual influences, promoting fruitful interaction, narrating and analyzing fictional scenarios, together they serve as a reaction vessel for inspired new ideas in science, technology, and beyond.

Whether fiction, fact, or forever undecidable: the Springer Series "Science and Fiction" intends to go where no one has gone before!

Its largely non-technical books take several different approaches. Journey with their authors as they

- Indulge in science speculation—describing intriguing, plausible yet unproven ideas;
- Exploit science fiction for educational purposes and as a means of promoting critical thinking;
- Explore the interplay of science and science fiction—throughout the history of the genre and looking ahead;
- Delve into related topics including, but not limited to: science as a creative process, the limits of science, interplay of literature and knowledge;
- Tell fictional short stories built around well-defined scientific ideas, with a supplement summarizing the science underlying the plot.

Readers can look forward to a broad range of topics, as intriguing as they are important. Here just a few by way of illustration:

- Time travel, superluminal travel, wormholes, teleportation
- Extraterrestrial intelligence and alien civilizations
- Artificial intelligence, planetary brains, the universe as a computer, simulated worlds
- Non-anthropocentric viewpoints
- Synthetic biology, genetic engineering, developing nanotechnologies
- Eco/infrastructure/meteorite-impact disaster scenarios
- Future scenarios, transhumanism, posthumanism, intelligence explosion
- Virtual worlds, cyberspace dramas
- Consciousness and mind manipulation

Giancarlo Genta

The Hunter

A Scientific Novel

Giancarlo Genta
Department of Mechanics
Politecnico (Technical University) di Torino
Torino
Italy

The persons, characters, events and firms depicted in the fictional part of this work are fictitious. No similarity to actual persons, living or dead, or to actual events or firms is intended or should be inferred. While the advice and information in the science part of this work are believed to be true and accurate at the date of publication, neither the authors nor the editors nor the publisher can accept any legal responsibility for any errors or omissions that may be made. The publisher makes no warranty, express or implied, and accepts no liability with respect to the material contained in either science or fiction parts of the work.

ISSN 2197-1188 ISSN 2197-1196 (electronic)
ISBN 978-3-319-02059-4 ISBN 978-3-319-02060-0 (eBook)
DOI 10.1007/978-3-319-02060-0
Springer Cham Heidelberg New York Dordrecht London

Library of Congress Control Number: 2013946546

© Springer International Publishing Switzerland 2014

This work is subject to copyright. All rights are reserved by the Publisher, whether the whole or part of the material is concerned, specifically the rights of translation, reprinting, reuse of illustrations, recitation, broadcasting, reproduction on microfilms or in any other physical way, and transmission or information storage and retrieval, electronic adaptation, computer software, or by similar or dissimilar methodology now known or hereafter developed. Exempted from this legal reservation are brief excerpts in connection with reviews or scholarly analysis or material supplied specifically for the purpose of being entered and executed on a computer system, for exclusive use by the purchaser of the work. Duplication of this publication or parts thereof is permitted only under the provisions of the Copyright Law of the Publisher's location, in its current version, and permission for use must always be obtained from Springer. Permissions for use may be obtained through RightsLink at the Copyright Clearance Center. Violations are liable to prosecution under the respective Copyright Law.

The use of general descriptive names, registered names, trademarks, service marks, etc. in this publication does not imply, even in the absence of a specific statement, that such names are exempt from the relevant protective laws and regulations and therefore free for general use.

While the advice and information in this book are believed to be true and accurate at the date of publication, neither the authors nor the editors nor the publisher can accept any legal responsibility for any errors or omissions that may be made. The publisher makes no warranty, express or implied, with respect to the material contained herein.

Cover illustration: 3D model of futuristic alien space ship in interstellar deep space travel. Copyright by AND Inc /Shutterstock.

Printed on acid-free paper

Springer is part of Springer Science+Business Media (www.springer.com)

Preface

Humankind is on the verge of epoch-making changes: as Tsiolkovsky would say, human beings are leaving their cradle, the Earth, to build a spacefaring civilization. We cannot say when humans will return to the Moon, land on Mars or launch the first interstellar spaceship; it may be just a matter of years or of centuries (particularly for the last enterprise), but one thing is certain: our development cannot continue indefinitely on a single, overpopulated and overexploited, planet.

Space is a harsh, dangerous and new environment, and new technologies must be developed to venture across its gulfs. This is however nothing new: Humans have been playing this game since the early Stone Age, when they left the plains of east Africa, the only environment in which they could live without the need to develop increasingly advanced technologies.

The new environments encountered and the new technologies required will, in turn, compel the human species to change, to develop new skills and new behaviours, on top of the basic human background that has characterized us ever since we first appeared on the surface of this planet.

Science-fiction novels are (in practical terms) experiments: the author creates a new environment and puts his characters in it to see how they react to the new challenges, and how they build their lives or succumb to the overwhelming stresses of life. In this novel, the action develops at the frontier of the tiny zone of our galaxy that has been colonized by humankind, where the few people who venture so far from Earth live in artificial habitats—space stations, mining colonies on asteroids and space ships—owing to the lack of naturally inhabitable, or already terraformed, planets.

In this situation, humans come in contact with something alien; not living creatures with whom they could have a relationship of some sort—friendship, understanding, hostility, hate or even just indifference—but machines that have presumably been sown into the galaxy by some intelligent species that intended to use them to explore and colonize its neighbouring stellar systems. This strategy—which had already been proposed by some scientists in the 20th Century—leads to a disaster because of the tendency by these self replicating machines to undergo evolutionary processes in an almost Darwinian way.

The confrontation with these alien machines causes a disruption to the usual standards of life in the affected zones at the frontier, and the characters have to face a situation that deteriorates progressively. While facing the new dangers and difficulties, they try to understand the limitations of artificial intelligence, a still controversial subject, in an effort to assess whether really intelligent machines might, after all, be possible.

The story is followed by a short appendix, summarizing the scientific and technological facts, theories and hypotheses that are behind the novel. It is subdivided into three sections: space travel, astrobiology, artificial intelligence and robotics. This feature is a characteristics of this new *Science and Fiction* series that has been introduced by Springer.

The section on space travel is the most hypothetical, because it is based on ideas which have, up to now, received no theoretical or experimental confirmation: A way of allowing the characters to overcome the speed-of-light barrier had to be devised and the warp drive approach was chosen. The astrobiological part is consistent with what is today considered most likely, whereas, for artificial intelligence and robotics, the books reflects my strongly sceptical views about strong A.I.

The author wishes to express his sincere thanks to Clive Horwood, the publisher of Praxis Publications, who produced my book *Lonely Minds in the Universe*, on the search for extraterrestrial intelligence, under the Springer/Praxis label; to Chris Caron (publishing editor at Springer), Stephen Webb (Science and Fiction book series editor) and Storm Dunlop (language editor) for their constructive criticism and suggestions, which resulted in a great improvement to the present text. The gratitude of the author goes also to his wife Franca, for the editing work she performed, as usual, on both the novel and the scientific commentary.

Giancarlo Genta

Contents

Part I
The Novel 1

1 The Hunter 3

Part II
The Science Behind the Fiction 111

2 Humankind on the Verge of Becoming a Spacefaring Civilization 113

Part I

The Novel

The Hunter

1

The corridors of the space station were empty, dimly lit by the night lights. Mike Edwards, second-class technician from the maintenance department, was walking slowly, trying to reach his quarters. He realized he was stumbling: twelve hours working outside were clearly too much. 'Safety rules would not allow shifts of this kind in any civilized place', he thought. But he was there for exactly that reason. A boy like him would never reach the status of second-class technician in one of those civilized places. And, above all, he was there because it was exactly in places like this that he had a chance to meet the Hunter.

Moreover, now that he had met Ann, this forgotten space station in the middle of nowhere could turn out to be heaven. Or hell, because there was still that bloody Madame … He didn't even know her name. Everybody simply called her that.

He was now in the private quarters section, located in the outer part of the huge rotating ring that provided an artificial gravity of a sort. Not much of a gravity, actually, but at any rate better than the weightlessness of the smaller space stations. Because true artificial gravity required too much energy, it was used only in starships. The floor was clearly curved, and your eyes were always telling you that you were walking upwards, while your equilibrium organs were telling you that the floor was flat.

The doors of the small rooms for the employees were set along the side walls of the corridor. In many places the paint was flaking off, and many of the doors were half open: empty rooms, not used for years. It was months since Mike had stopped being aware of the derelict look of the living quarters on the station, too big for the number of people that now lived there. Actually Mike was too tired to think about such things. He was walking like a robot, eager only to get to his room and to throw himself on his bed. Suddenly he realized he was there. With what felt to be the last of his strength, he raised his right hand and pressed his thumb on the scanner.

"Open!", he whispered, hoping that the lock could recognize his voice. He heard a low squeal and thought 'I should find time to fix that door, before I get locked out of my room'. Finally he pushed open the door and, without even looking around, he let himself fall onto his bed.

He ordered the door to close and realized he had not even the strength to undress and take a shower. (Another safety rule overlooked: you were not supposed to leave the airlock without showering and, above all, without changing the coveralls worn under the space suit.) He laid on the bed ready to sleep, still in his dirty gray coveralls, covered in oil and sweat.

'To hell with the alarm clock. After a shift like that, tomorrow I will be late for work'. He realized he had promised Steve to take a look at his waiter robot, but he decided that the robot could wait too.

He was almost asleep when he noticed a small blue light blinking under his communication screen. 'No, not now', he thought, Even so, he ordered: "Message on the main screen".

The screen came on:

New Shanghai space station
Internal communication network
Message recorded by traffic control room—urgent.

Under those few lines there was the signature of the acting traffic controller, Joe Ivanovich.

Mike jumped out of bed and sat on the edge. The sudden surge of adrenalin made him oblivious of the tiredness his body was still feeling. There was just one reason for a message like that from Joe: a simple message saying nothing, but that could cost both of them their jobs. Actually the charge would be one of giving out classified information for Joe, and one of corruption for himself—who would believe that Joe would run such a risk just for the sake of their old friendship?. A traffic controller could do nothing worse than giving out the movements of the ships that arrived or departed from the station.

He opened the case where he stored the tools he used to fix his friends' robots, and took out a small metal box with two connectors trailing from it. He connected one to the output of the communication screen and the other one to his small portable screen. The screen came on, and a single line appeared: *Your friend will be here for dinner at 18.00.* He was really proud of that duplicated encryption. The message was coded in hardware by that innocent looking interface, but even if someone could decode it, only something that was still apparently uncompromising would appear. And, finally, even if somebody could understand who the 'friend' was, the time would tell them nothing, because no ship would dock at 18.00.

The *Morning Star* would dock eight hours earlier, at 10 a.m. As soon as his heartbeat returned to a normal rate, he realized that it was not as lucky as he had thought at first. He looked at his watch: five a.m. He had just five hours. How could he get some rest and organize what he had planned in just five hours? It was impossible, absolutely impossible. He fell back on his bed. He could almost have started crying, more out of tiredness and anger, than from disappointment. He had been there for almost a year, just waiting for this moment. He had prepared every detail with a view to meeting the Hunter; he had rehearsed and re-rehearsed a hundred times the words he would say. And now he also had something of value to offer him, a piece of information that could make the difference. But everything was useless if he was deeply asleep when the Hunter left his ship.

He reconsidered the situation. The *Morning Star* would dock at 10. The last checks would take 2 h, the antimatter transfer could not start before half past noon and the Hunter would not leave his ship before it was strictly essential. For years he had studied the man's habits and he was sure he would behave as usual tomorrow. He would not reach Steve's canteen before one: he had three hours more to get ready. It was not much, but he could manage.

Now he had a decision to take. He had promised Ann to let her know when the Hunter was due to arrive. And if he didn't call her Madame would certainly get very upset. And what about that? What if she wouldn't let him work on her RGs? No, he had to call.

"Call Ann, private address", he said in a loud voice.

A few seconds went by slowly, then a sleepy face, sticking out from under a blanket pulled up to her chin, appeared on the screen. Certainly he couldn't say that under these conditions she was beautiful, but he remained there, speechless, contemplating her face.

"… the hell calling at this time? What's the matter?", was all he caught as soon he managed to focus his attention again. And suddenly he was aware of two things. First, it was five a.m.; second, she had answered and, above all, she had not switched off the video. He was starting to speculate on the possible implications, when he realized he had to answer her.

"I've learned that the Hunter will be here tomorrow morning. I think that he and his crew will be at Steve's for lunch at about one. Tell Madame to get her RGs well spruced up and perfumed, ready for action", he answered hurriedly.

"Thanks, Mike", she answered with a smile. "We'll all be at Steve's at one …".

That 'we' spoiled all the pleasure her smile had caused.

"Please, tell Madame to go easy and, above all, not to have them wear those dresses … And instruct the RGs not to try anything with him. The Hunter is known to be a puritan, and since his wife died in that accident with his fifth replicator, he has never looked at a woman ...", he answered.

"All right, your idol is above anything material. He has never looked at a woman, and soon he will grow a pair of angel's wings. I will look out for a small statue so that you can make an altar in your room and worship him as he deserves ..." she answered with evident sarcasm. "We will make sure we tempt your idol suitably".

"No, don't say that. And, above all, don't distract him. He has to listen to me. Tell Madame to instruct her RGs to deal with his crew. The first mate should be a good target: they say that he spends most of his share of the rewards on that kind of thing. It is a pity I could not finish that work on RG46B/G, but anyway she should work well enough to siphon a good quantity of money from the first mate's pockets into Madame's ..."

"All right, you mean that whore called Lulu. There's no point in you going on calling her by that official designation ... Very professional, no doubt, but we all know what you do with her".

Mike refrained from smiling. He wanted to tell her the actual reason he spent all that time working on that robot, but obviously he couldn't. But why that reaction? Was she jealous? Perhaps he should say more.

"Well, that piece of junk is no longer as she was when came from the factory. But where did Madame get her RGs? You cannot ask me to re-program a robot without testing it, can you?"

"If you want to test a robot, get Steve's waiter. At any rate thanks for the tip. We will all be ready at one."

Yes, undoubtedly Ann was jealous of that robot. Mike pulled up his blanket and tried to sleep. But he was too excited: The Hunter was due to arrive soon; moreover, Ann was showing some interest in him.

It was a pity he could not finish his work on that RG46. When he had suggested to Madame that he should improve her program, the robot had a voice that was not as sexy as that of an emergency beacon. Now, on the other hand ... He had downloaded tens of porno videos and put the voices of the actresses through a spectrum analyzer, then used those spectra to re-program her voice synthesizer. For days he had spent all his free time on that job, and now the results were impressive. Then he had started working on the controls for the actuators that allowed the robot to move. And now she was able to perform a striptease like a high-class dancer.

He remembered the old joke about striptease: do not try to understand how they do it, just enjoy the show. He didn't enjoy any show, but on the contrary he had analyzed every detail of her movements, and had re-programmed all the amplifiers powering the tens of electrohydraulic actuators that gave life to those seventy kilos of machinery that was covered with synthetic flesh and skin. And now that robot could be the star of a first-class night club on Earth, instead of performing in a brothel on a space station in the middle of nowhere.

Another few weeks of work and Madame would have no reason to ask Ann to do that job. RG46 would have been better that any flesh and blood girl: That was the aim of all this work. It was a real pity he could not finish his job before the Hunter arrived.

That thought brought him back to the present. He needed badly to sleep, otherwise tomorrow he would miss the opportunity for which he had come all that way out there. He opened the small cabinet where he stored the few prescriptions he sometimes used, and took one of the pills Joe had given him a few months ago, with the recommendation to use them only in case of absolute need: the pills astronauts used when they needed to get some rest in an emergency. He looked at his watch: it was forty minutes past five. He set the pill on four hours and a half, and swallowed it.

His head did not even touch the pillow before he was asleep: a heavy and dreamless sleep.

2

New Shanghai was a large space station located at the L5 Lagrange point of the fifth planet orbiting the main component of the double star BD -05 1844, 9.2 parsecs from the Sun. The red-orange star, slightly smaller than the Sun, was known well before humans started their expansion in space, even if it couldn't be seen from Earth with the naked eye. It could be found in star catalogs under various names, such as BD -05 1844 or Gliese 250. Fifty years earlier the Shaanxi Terraforming and Space Engineering Corporation had the space station built in the shipyards of 40 Eridani. Towing it to its final position had been considered a technological miracle: four deep space tugboats propelled by warpdrive, had worked for two years to tow it the 6.2 parsecs separating the two star systems. Actually it had been an epic feat, that was, however, only a part of a much more complex project. The space station was meant to be the logistic headquarters for the terraforming operations to be performed on the star's second planet: a planet slightly larger than Earth, located right in the middle of the habitable zone.

Owing to the complete absence of lifeforms, the planet had an atmosphere made of carbon dioxide and nitrogen, but could become perfectly suited for human life with relatively simple terraforming operations. It had some large oceans, mountains, rivers and lakes and a water cycle similar to that on Earth. Once terraformed, it would host a few billion human beings, becoming the most important planet in that part of the Galaxy. The STSE Corp. had obtained all rights to the planet and had launched a large-scale colonization operation. The hopes were so high that the planet was even provisionally designed 'New Earth' and provided with a space elevator. The first stage of the operation had been to build some huge nuclear power stations orbiting the fifth planet, powered by deuterium and helium 3 mined from the atmosphere of the gas giant. The energy thus produced was used to produce antimatter, which was then stored in the tanks on New Shanghai, where the thousands of starships participating in the project could be re-fuelled. Other space stations were due to be built, and the value of the shares of the STSE Corp. were soaring in the stock markets on Earth and on the largest colonized planets.

At the same time a rocky satellite of the third planet of the small component of BD −05 1844, a red dwarf, was discovered. The satellite had the peculiarity of hosting some primitive lifeforms, nothing more than bacteria, that through the millennia had transformed its atmosphere, enriching it in oxygen and making it breathable. The satellite, named Ceres owing to its potential for agriculture, had been immediately colonized and, over time, came to host a population of a few millions. They too were employed in the terraforming operations on the main planet, mostly supplying agricultural products.

The system seemed to be bound to become one of the main centers in the outer parts of the galactic zone colonized by humankind, when suddenly, as often happens in human enterprises, disaster struck. In this case the disaster took the form of some strange objects, apparently huge self-replicating robots of alien origin, which attacked a few human colonies at the frontier of the inhabited zone towards the Monoceros constellation, where the New Shanghai station was located. Human expansion in that zone faltered and the value of real estate sank. Following this crisis, the companies operating in the terraforming sector, like the STSE Corp, had a severe setback. Only the direct intervention of the Chinese government avoided the bankrupcy of a company that just a few month earlier seemed to be booming. The situation had been stagnating for twenty years and, although operations on the planet never completely stopped, nobody now called New Earth anything more than just a derelict planet, with little prospect of being terraformed in the predictable future.

Slowly the colonists who had settled Ceres also started to leave and now the colony had slightly more than 30,000 inhabitants. Most of the settlements were now nothing more than ghost towns.

New Shanghai was now reduced to a lonely station at the periphery of the colonized zone, a harbor where the few starships that still ventured into that forgotten region of space could find antimatter and some assistance.

3

Mike woke up with a terrible headache. He looked at his watch and realized it was ten past ten: those pills worked like a precision clock, although he could not tell whether they were free from side effects.

If Joe was right, the *Morning Star* had just docked. Mike switched on the monitor to check the arrival schedule: it was showing that a private ship was completing the docking procedure. Nothing strange in the Hunter asking for the name of his ship not to appear on the list: he was too famous and preferred to have some privacy.

Mike took a quick shower, put on his best technician coveralls and started walking towards Steve's canteen.

He sat at a table, and within a few minutes the waiter robot came by to take his order. He asked for a low-calorie breakfast and added "Tell your master to come as soon as possible. I need to speak with him about an important matter."

Not a minute later Steve came out from the back of the shop where the kitchen was located and sat down in front of him.

"What's going on? And why are you here at this time? Maintenance technicians on strike today?", he asked.

"Never mind about the time, I need your help. The Hunter has just docked and …" Mike started to say.

"If you are here for what I think you are here for, the answer is no", Steve interrupted him. He had immediately understood what kind of help Mike was asking for. There was little guesswork needed: together and with Joe they had discussed how to talk to the Hunter tens of times.

"But why? We've discussed it so many times, we've ironed out all the details … And now that he is here, you just pull back?", Mike reacted, siezing his wrist. He was not prepared for a reaction like that.

"Listen, Mike, it's one thing to talk when you think these things will never happen, it's quite another allowing yourself to get into trouble. Or rather, to get all of us into trouble. Joe was wrong to tell you he was arriving: he's risking his job. And me too. If it comes out that I allowed you to disturb an important person in my canteen, I could have my license taken away. And you too can get fired. Wait, if he is here to have his ship fuelled up, he will spend at least 24 h in this station. When they give the news of his arrival, you can send him a written request for a meeting, and everything will be above board."

"You know full well that won't work … As soon as the news of his arrival is made public, his terminal will be jammed with emails. And he will read none of them. And if I miss this chance …". Initially he had been able to remain calm, but now he was getting really angry.

"Don't do it like this, Mike. You know that it would be better my way. For all of us. Dreams are one thing, but the real world is another."

At these words Mike was unable to restrain himself any more. Squeezing Steve's wrist, he interrupted him: "Nice friend you are. Think of all the hours I've spent fixing that robot of yours that is falling to pieces. Now I can call it and disable it with a high frequency discharge. Instead of pretending you have a broken waiter, you will have to play the waiter for a week …"

"But what do you think you can gain in this way? You cannot just get to him, saying that you know where a replicator is hiding and offering to tell him the coordinates if he takes you on his ship. It's nonsense, he will never believe you."

"Leave that to me. I've rehearsed tens of times what I'm going to tell him, and I will be able to make him believe me. Only help me to get close to him, and I'll do what's needed". Mike was a bit less excited now and realized that perhaps he could still convince his friend.

"But you cannot stay here the whole day. It's late, and you should be at work by now …"

He could easily get around that objection. And, once this marginal point was overcome, perhaps he could convince him. "Yesterday I worked outside adjusting the long-range warning system for twelve full hours and today I deserve a whole day off. And then, if all goes well, tomorrow I'll resign and say goodbye to this rathole. Please, let me try".

Steve shook his head. "Do whatever the hell you want, but I know nothing about it. You can disable my waiter. Now I'm going, do as you wish" he concluded, with a skeptical expression.

Mike felt he had heaven within his grasp. "No problem, if we get into trouble, I will take all the blame. And I will enable your waiter again as soon as this is over". He stopped to think for a few moments, and then added: "and next time I have RG46 to do a full rehearsal, I will tell you, so you will enjoy a striptease that you can see only on Earth".

"All right, don't forget it. I am sure that tomorrow you will still be here. This thing will never work." He finished by getting up and going back to the kitchen.

Mike called the waiter. As soon as the robot got close, he switched on the tiny high-frequency generator he had in his pocket, and the machine froze. He got up suddenly and went to knock at the door of Steve's kitchen. "Steve, your waiter's frozen, I think a power amplifier has burnt out and it lost its power. I've no time now to fix it, shall I bring it in?", he said in a loud voice. He saw that two technicians were walking along the corridor and he wanted them to be able to testify that the waiter had a problem, just in case. He hoped it was a needless precaution, but you never knew.

Steve came out. "No, not again? That's the third time in a month. Why can't you fix it better? And now, what can I do if some customer shows up for lunch?" His desperate expression was almost credible. At least as far as one of the two technicians, that now were quite close, was concerned, because he told him: "You can't blame Mike, he knows how to deal with robots. You should get a new waiter, can't you see that this one is falling into pieces?"

"Yes, and where do I get the money? When you come here to fill up with beer I'll just double the price, so that I can pay for a new waiter. Can't you see that this hole is falling apart?" he answered with an unhappy face. "And now what can I do?"

"Don't worry, if you need help, I am here", Mike assured him.

They pushed the robot into Steve's room without saying a word and Mike went back to his table. He took out his small screen from his pocket and started reading the news. Now the only thing he could do was wait. At one, he saw Madame coming along the corridor with her four RGs and Ann. As usual she was elegantly dressed, and the RGs had on their working dresses … if they could really be called dresses. And Ann had very little more on. Just as if he had not warned Ann not to exaggerate what the girls were wearing … And then for Ann too, to go around dressed like that, like a whore! 'Well, and what is she, after all?' he thought. 'No, it's Madame that forces her to do this', he concluded, trying to keep his sense of proportion.

For a moment he thought of getting up and telling them to go away and change their dresses, but then he decided it was useless. If Madame had decided to have them dressed this way, he would not be able to get her to change her mind by making a scene.

For a while he watched how RG46 was walking—he continued to refuse to call her Lulu—and compared her movements with those of the other RGs. There was no comparison, he had done a terrific job. If he just could carry on with it, Madame would definitely not need Ann any more.

He suddenly realized that there was something wrong: how could he reprogram those RGs if he was leaving with the Hunter? Only now he realized that his goals were in conflict with one another. But he could not abandon his dreams so soon. And then, he had always thought that if he could participate in the hunt and have a role in destroying a replicator, he would earn enough money to take Ann away from this place, get her free from Madame, and perhaps even marry her.

Thinking about the Hunter brought him back to the real world. It was already a quarter past one and he had not yet shown up. Was it possible that Joe was wrong and the arriving ship had nothing to do with him? There were a lot of people who preferred to keep their movements secret.

He sat down again, trying to keep calm. From time to time he threw a quick glance at the door leading to the docking zone, and at the six women—or rather two actual women and four robots—sitting about twenty meters from him, beyond the canteen area, on the sofas of a small waiting room just off the corridor leading towards the private quarters.

At half past one he was so nervous that he got up and went to the door of Steve's kitchen. He knocked and asked Steve for a drink. "If you hadn't done your nice job on my waiter you would not have had to come in person" Steve said. And then he added: "I see that Madame and her whores are all here, ready to hunt for the Hunter. Artificial whores and natural whores, I mean".

Mike pretended not to understand. He knew very well what his friends thought about Ann, but he didn't want to start the endless discussions on the subject yet again. And he knew quite well that they thought that the main reason he wanted to go away with the Hunter was to get away from her—and hopefully to forget her.

"At any rate your Lulu is gorgeous, you've made a wonderful job of her. I dare say that the only human in the group looks more like a cheap RG than her", he ended.

Again Mike pretended not to hear, and went back to his table, resisting the impulse to answer him as he deserved. It was useless to repeat for the hundredth time that if Ann was there it was only Madame's fault, and if he could only get her away from the station everything would be fine. He knew full well that Steve and Joe had taken on a sort of crusade to make him forget Ann. For his own good, obviously, but that upset him even more.

Suddenly, at about two, the door opened and a group of people came in from the corridor. He looked at them one by one: he had seen their pictures so many times that he recognized all of them. Leading the group there was Andrei Romero, the first mate. The second one was Ali O'Connor, the chief engineer, and then, one by one, all the others. The last one to come in was the Hunter himself. He realized that he was staring at him. For years he had been dreaming of meeting that man, the man who had destroyed six replicators, those awful alien machines that were bringing havoc and death to the colonized worlds.

Soon the group was in the area where the corridor widened to become the canteen area. They all sat round a long rectangular table. There were eighteen of them, but there was enough room for all of them, and some space to spare. Mike noticed that many were taking quick glances at the small room where the RGs—and unfortunately Ann as well—were sitting, giving what he thought was an indecent show. Romero had managed to sit in a position from which he had an unobstructed view in that direction. 'Then it's true what they say about him', Mike thought. 'Madame will be happy. Now it is up to me to have him make the best choice'.

He got up, and took the eighteen menus he had prepared. He came up to the table and, trying to behave as he had seen human waiters behaving in old movies, before robots had completely taken over those menial jobs, he gave them the paper sheets.

"Eh, boy, since when did they have a technician, and nothing less than a second-class technician, serving at table? Don't they have robot waiters in this hole? Or is this perhaps your hobby?", Romero said to him in a derisory tone.

"I am sorry, sir, but the owner of this place had a problem with his waiter, and since he is a friend of mine I am trying to help him get over this emergency", Mike answered, blushing.

"And you are a maintenance technician? Fix his robot, instead of acting the clown like this", the first mate went on.

"Sorry, sir, but my specialization is long-range warning systems. Working on robots is none of my business", answered Mike, hoping he would not realize that he was lying. "The waiter was sent to the robot maintenance section, but will not be operational until tonight." Things were turning out badly. He hadn't expected a reaction like that.

"Stop it, Romero", the Hunter cut him short. "Don't you see you are embarrassing the boy, who after all is here only to help a friend, and to allow us to have a good lunch?". And then he continued, addressing Mike with a smile "Don't listen to him, boy. We understand the problem, and we thank you for your help. After all, it is not something that happens often to have a technician, and a second-class technician at that, bringing us food". "Thank you, Sir. I will try to do everything properly", Mike answered, starting to take the orders. Luckily that emergency looked to be over.

He served the appetizers, without any problem. Then it was time for the main course. Every time he got close to the Hunter he tried to start the short speech he had prepared, but he didn't manage to do so. There was little he could do about it, because the man made him feel ill at ease.

Every time he went out to the kitchen to get the food, Steve asked him how things were going.

When he went in to get the main course, Steve scolded him: "If you do not start, you made all this mess just for nothing. At least try ... the worst thing that could happen is being sent to hell. If you don't even try ..."

"I just cannot. I am too nervous. Perhaps tonight at dinner ..."

"No, don't even think about it. For tonight my waiter must be ready. Go on."

When he got out he was ready to go on with his self-appointed mission—but he left the Hunter to be served last.

As soon as he put a dish in front of Romero, the first mate stopped him: "I see you have a good number of those creatures that polite people call robogirls and that less polite people like me call in a lot of different ways." He stopped to wait for the others to stop laughing. Then he went on "It looks like one of them is an RG46, three are older models, but I don't understand what the hell the fifth one is".

Mike wondered how he could tell the model from such a distance. He was obviously a true expert, so there was little to say except "Yes sir, that one is an RG46/G …".

"Obviously, one can't expect that in a hole in the middle of nowhere like this they have anything other than general purpose RGs. But I am interested in the fifth one."

"The last one is a real girl. A human, I mean", he had to admit.

"No … who would ever expect that in a place like this they had a real girl. The stop here may turn out more interesting than we expected."

Mike was desperate. He glanced at the Hunter, who was looking at Romero with clear disproval. He had to do something before the Hunter cut short the subject and the first mate made a decision about the human girl. "If you allow me, sir, I can tell you that this RG46 is quite a special one …"

"Don't try to convince me, boy. I have no desire to go to bed with an emergency beacon …" the first mate interrupted him.

'He really is an expert', Mike thought while everybody else was laughing.

"You are perfectly right, sir. The voice synthesizer of the RG46 model is awful. At least emergency beacons try to put some warmth in their voice to reassure stranded astronauts", he tried to joke. "But as I said, that one is quite special: I spent months re-programming her and now … Listen to her, and you will tell for yourself. And it is not only the synthesizer. I have re-programmed her from scratch. Ask her to perform a striptease, and you will have a surprise. I can tell you, she is much better than that human, who, between us, is not that much. And then it will cost you a quarter as much. Take my advice, the quality/price ratio is much, much better. But, please, don't tell Madame I told you this". He realized he had gone too far, and suddenly stopped speaking.

He glanced round and realized they were all laughing. All except the Hunter, who had a suspicious and disapproving look.

"Come here, boy", he said in a voce that allowed him no option.

As soon as he was close to him, he went on: "And so you have programmed that RG from scratch". He pronounced the acronym with utter distaste. "Voice synthesizer, actuators and all the rest, but you are not able to reactivate a waiter that stopped working?" he asked looking at him straight in his eyes.

"No, sir. Software is my specialty, but I am not able to work on a robot's hardware, and that waiter has a burnt out component" he started to justify himself, realizing just as he said it that the excuse was not working. The hardware of waiter robots was so simple that what he had said was clearly nonsense. And then he suddenly broke his reserve. "Please, forgive me, sir. That story of the broken waiter was just an excuse to have a chance of talking to you ..." he confessed in a low voice.

At these words two of the men suddenly jumped to their feet, went behind him and seized his wrists, locking him in a painful grip.

"I'm glad that you are not one of these fanatics who would like to get close to me to stop me—forever", the Hunter answered him, in a calm voice. Then he spoke to the two who were still holding him: "Don't worry, don't let this good lunch get cold, leave him and get back to your food". The two exchanged a disapproving glance and went back to their seats.

"Good heavens, boy, you could have written me a message. Why all this comedy"? "I was afraid that you wouldn't read it. Perhaps you do not realize how famous you are. I guess you receive thousands of emails every day ... how can you read them all?"

The Hunter realized that he had a point. "Then tell me what you have to say, and then go and fix that waiter, if it was ever actually broken". Mike was sweating and feeling much worse than he had the previous night, even after all those hours in a space suit outside the station.

"Sir, I believe I have a good idea of the position of a replicator", he said in a single breath.

The crew looked to each other in disbelief. They had heard such statements hundreds of times before: Ever since hunters had started looking for replicators, trying to understand their moves or to get their positions had become a common occupation.

The Hunter refrained from the impulse to send him away, but then he thought that if he had gone to all the trouble to prepare this comedy, he might actually have some good reason.

"Sit down and calm yourself. Then tell me about this replicator of yours", he finally said.

4

"About twenty days ago", Mike started, "a patrol ship, cruising in the mining area of the asteroid belt that lies where the gravitational zones of the two stars in this system merge into one another, found a battered ship, with an old miner on board who was in a state of severe shock. The ship was towed here and the miner was brought to a better-equipped hospital. The poor guy kept repeating that he had been attacked by a replicator that intended to land on the asteroid that he was mining."

"The first thing you must remember is never to believe old miners, neither when they boast of the immense wealth they have found, nor when they speak of attacks from pirates, raiders, replicators and so on that they only just managed to escape.", the Hunter remarked.

"You are perfectly right, sir, but this one was not drunk. He was deeply shocked, and went on and on repeating the same story. And then I had a good look at the hull of his ship, and I am sure that the damage could not be due to normal weapons. It looked as if it had been hit by plasma jets, very powerful jets, too" He paused for a few seconds, then went on "like the plasma jets they use to cut scrap metal, but much, much more powerful".

"Were the drops of fused metal located on the inner or the outer surface of the melted edges?", the Hunter asked.

"There were no drops, Sir. It looked as if the metal had vaporized, not melted".

This was a point for the boy. It looked as if he had actually seen the wreck of a ship attacked by a replicator. Either that or he had read quite detailed descriptions of the effects. "Is it possible to see the wreck?", he asked.

"Unfortunately not, they scrapped it. But I have hundreds of pictures, which I can download to your computer. And you can compare them with the pictures you can find in this space station's archives; they are less detailed, but they come with a full, and official, documentation. So you can be sure I am not trying to sell you pictures found on the net …"

"And where did this miner meet the replicator? Do you have detailed documentation, so that we can find that asteroid?"

"Yes, sir, I copied that ship's log and I can find that asteroid at any time".

"Can you explain how that miner was able to survive this encounter?", Romero interrupted him. "You are trying to make us believe that a replicator attacked a small mining ship and then the ship was able to escape, even if in quite a bad state? Do you think we are so stupid as to buy all that?"

Mike turned to face him. "I have no explanation, Sir. But that miner had stopped here six months ago and asked me to adjust the long-range sensors of his ship to increase their range. From the ship's log you get the impression that he detected the replicator when it was still far away, and that he started running immediately. When the replicator arrived, it fired just two or three shots, but then it didn't go after him. I have no idea why it behaved like that".

"Nonsense", Romero concluded in a tone that admitted no reply. "Replicators never settle on that. They go on shooting until their target is completely destroyed".

"Yes, but there could be an exception", the Hunter interrupted him. "If that replicator had already started the replication sequence before realizing that someone was on that asteroid, it could not go after him. It had no choice other than mooring to the asteroid and continuing with the replication process". Mike looked round and realized that not all the crew were convinced by that explanation, but nobody felt like contradicting the Hunter. "I didn't think of that, Sir, but it could explain what happened", he said. Then he realized what that implied. "But then ..."

"Yes," the Hunter confirmed. "When did that encounter take place?" he asked.

Mike took a sheet of paper from his pocket. "I believe it was fifty six days ago, Sir".

"Then we have no time to waste. Either we get it within ten days, or there will be two replicators in this system. Give me the coordinates, and tell the maintenance people to hurry up with the refueling operations".

Mike tried to muster up all his courage. "Sir, before giving you the coordinates, we must discuss ... my conditions".

The Hunter grimaced. "You must know that if your tip is correct you will get one tenth of one percent of the reward. Don't you trust me?"

Mike stopped for a moment to think what he could do with that huge amount of money. "Sir, I am not interested in that. What I would ask you is for me to be enrolled on your ship".

Everybody started laughing. "We don't need waiters on board", Romero said. "Not even programmers for RGs. Even if, after all, you bring that RG46– provided that what you told us is not nonsense—would we take you with us".

The Hunter pretended not to have heard his first mate's words, and looked Mike straight in the eye. He was said to have an instinct for judging people and he felt he could trust this boy. "Well, we could certainly use a maintenance technician, but not a second-class one. We could take you as a fourth-class technician, on the minimum wage and obviously with your share of any rewards we get while you are on board. That's after we have checked that what you told us is not a bunch of lies."

Mike was delighted. "You will never regret this decision, Sir. Give me a dedicated area on your computer, so that I can download all the relevant documentation." He paused for a moment, and then added: "Is it really so important that we leave as soon as possible?"

"It is vital that we get there before the conclusion of the replication process. But here you take 24 h to transfer the antimatter on board. And then there are the usual maintenance operations."

"As for the transfer process there is no problem, we have still the old equipment, but now you have another maintenance technician. With your permission, I can inspect the ship and prepare all the spare parts, and then I can do the maintenance operations after leaving". And he thought 'This way we can get out of here as soon as possible, and if Romero follows my suggestion, he will have no more time for Ann ...'

His thoughts were interrupted by O'Connor. "You are crazy. Do you mean you want to go on board during the antimatter loading process?" "There is no danger, Sir", Mike answered. "Only the new fast loading equipment requires those new safety precautions. Previously it was customary to remain onboard to inspect the ship while loading the antimatter. There is no danger at all", he said again.

"Is that true?" the Hunter asked. He remembered something like that, but he was not sure.

"Yes. We are now so used to fast loading that we do not even remember the old methods, but actually we can get on board", O'Connor explained.

"Well, so we leave in 23 h. I want everybody to be strictly on time", the Hunter stated.

5

The following day Mike got up early, even though he had almost nothing to do. He had already carried on board all the spare parts he would need, together with the bag that contained all his belongings. He went to say farewell to Joe and Steve, and then went to the personnel bureau to resign from his job. They made some difficulty, but he felt they were not really unhappy. With the current situation, having one wage less to pay was not bad news. As soon as he had finished with that, he started toward Ann's room. He had called her the previous evening and she had invited him to come to see her before leaving.

He knocked and the door opened immediately. As soon as he entered he realized that she was wearing the same dress as she had on the previous day, when she was there with the RGs. He was so surprised that he was unable to speak for a moment.

"You don't like it?" she asked, closing the door.

"No, it's that …", he started to answer. Then he thought he understood. "So did Madame order you to repay me that way for the tip-off yesterday?", he said in a single breath, without really believing his own words.

"Stupid", she reacted, slapping him in the face. Then she clearly regretted what she had done, because she added, in a steadier voice: "I'm sorry. Does it hurt?"

"Yes, but I deserved it. Please, forgive me". Then, after a moment, he added: "It's that to see you there yesterday, with those RGs, dressed like that … I was really shocked".

"Well, as we are apologizing, you owe me some explanations. You must tell me what you told that first mate about Lulu … and about myself …", she went on in an angry voice.

Suddenly Mike blushed. "Uh … ya, now I'll tell you", he mumbled. Then he realized that he owed her an answer. "That first mate is really a connoisseur. Even from that distance he realized that Lu … that RG 46G is an RG 46. It's really incredible" he concluded with a slightly steadier voice.

"Well, sure, so he is a connoisseur. But what did you tell him about me?" she went on, without giving him time to prepare an answer.

Mike realized he was walking on a minefield. "He looked at you … and thought you too were an RG, but he was unable to tell what model. So I had to tell him you were human." He answered, more and more ill at ease.

"I hope he did look at me. Do I look like a person that can go unnoticed, particularly if I am dressed like this? But I think you also said something else about me …"

Mike looked at her for a moment. All the pleasure he felt looking at her was spoiled by the consciousness that she allowed so many people to look at her like that … and by his vain attempt to find a way out. "Yes, of course. Naturally he was staring at you. I didn't mean that. I just meant that he was trying to understand what kind of RG you …"

"Don't search for excuses. He told Lulu what you said …"

'I am done', he thought. All RGs recorded everything, word for word, movement by movement, even if their owner needed to cancel their memory within 24 h. It was simply to verify that the job done was exactly what the customer had paid for and, in case of any damage, to claim a refund. Anyway, he had heard of criminals who had been framed by their own words said to an RG in a moment of ecstasy. And now he was the one to be framed by the words of an RG.

He decided it was useless to look for excuses "I told him that the RG was much cheaper than you" … he started.

"And that after all I am not worth that much, isn't that so? And what about the quality/price ratio …", she interrupted him.

"All right, I told him all that. But you know that I did it to …" He stopped because he didn't want her to get any angrier than she already was. "But did that bloody RG tell Madame everything?", he asked.

"Obviously, she told her. I had nothing to do with it.", she answered.

'That's true', he thought. 'A robot answers only to its master'. And then he realized that Madame would not allow him to see her any more. "And how did she react?", he asked.

"And how should she react? She started laughing and hugged me … and in the end she had to admit that I was right", she answered, suddenly changing from the angry look she had been showing up to then into a smile.

He was astonished. "Sorry, but I am completely confused. Wasn't she furious? And what were you right about?"

"Yesterday she didn't want me to go out with the RGs: she said that those were not the usual station hands, without a penny to spare, who couldn't help but to settle for a robot. If one of them wanted me, she could not make him change his mind just by raising the price …" she started.

"But wasn't that exactly what she wanted?", he asked.

"Obviously not, stupid", Ann started. Then, seeing his confused expression, she added "Sorry, that's the second time I've called you stupid in a few minutes", she added.

"At least this time I didn't get slapped …" he answered feeling his still reddened cheek with his hand.

"Does it still hurt?" she asked in a subdued tone.

His grimace was more eloquent than words. She hugged him and kissed his face.

"If that's the procedure, next time you call me stupid I want the slap as well", he said. But then he went on "Anyway, it seems that I am really stupid, because I am still in the dark. Why were you right?"

"Because I had told her that there were no problems, that you would convince them that Lulu was much better. After all, with all the work you put in so that clients would choose her …"

"And so you realized that?" he asked her.

"Actually it is not that we realized it. Madame convinced you to work on Lulu to be sure that there was no danger that some client would ask for me", she concluded, smiling.

He was speechless. On the one hand he was overjoyed hearing this unexpected news, but on the other hand he felt he had been manipulated by the two women.

After a while she asked him: "But did you really think that Madame was using me that way … And that I was a whore?"

He raised his eyes and answered "I tried to deceive myself and pretend that it was otherwise, but all appearances suggested that was the situation. But then, why do you live with Madame?"

"What do you know about Madame?", she asked without answering his question.

Mike was quiet for a moment, then answered "When I think about it, I know almost nothing. I know only what people say about her"

"Well, then what do people say about Madame … and about me?", she interrupted.

"The general idea is that when she was younger she worked in a brothel on some important planet, some even say on Earth. You know, when she wants to, she can be so elegant … And that she must have been beautiful when she was young, clearly not a girl you would find in some rathole in the middle of nowhere. And then she was sent away, or she got away for some reason. The idea is that she stole those four RGs, to start her own business. About you, some say that you are her daughter, others that you are a girl that worked in the same brothel, and that you came away with her." He answered, choosing his words carefully.

"Do you want me to tell you how things really are?", Ann asked, without commenting on his words.

"I'd love you to", he answered, looking at his watch.

"I'll be quick … you still have time." She paused, as if she needed to collect her ideas, then went on: "Ruth Donovan, the lady you call Madame, was the head of the planetary biology lab of the helium-mining base on the second moon of the fourth planet of BD -03 2001. She had little to do there, since there was no trace of any lifeforms and her only job was to check that there was no biological contamination from outside".

"Madame was a biologist? I find that difficult to imagine", he interrupted her.

"My mother, who was the head of personnel on the base, asked her to keep an eye on the school. On the base there were about twenty boys and girls and a few babies, and my mother thought it was a nice idea for there to be a human looking after them, even though they had their robot teachers. I was in the last year of high school, and I planned to go to the University on 40 Eridani 3," she went on, without taking any notice of his interruption.

"How many years ago?", he asked.

"Seven years ago", she answered. "One morning I will never forget, when I was doing an astrodynamics exercise with my robot teacher, we were attacked. I suddenly heard some explosions and I found myself on the floor, with a heavy beam that had fallen on me, preventing me from moving …"

"Now I remember! The Helium-Mine Massacre!", he exclaimed. "But there were no survivors. How did you …"

"Let me go on", she said. "I remember seeing Madame entering the room where I was, all covered with dust. She had scratches everywhere, but was not seriously injured. She tried to move the beam to free my body, but she couldn't. The robot teacher was no longer working, so she told me not to worry: she would look for a robot to get me out of there. She ran outside. She knew that there was a night club close by and thought she could borrow two or three RGs. You know, the actuators of RGs are powerful, particularly those in the legs and the lower torso, and she hoped to get me free in that way. Actually, when she got out of the school, she realized that things were much worse than she had thought. There were corpses everywhere, and there were still explosions not far from where we were."

"She got into the night club.: there was the body of a man in the entrance, and an RG that was not only naked, but with the metal structure sticking out from the synthetic flesh. The synthetic flesh was completely detached at many points, but the structure and the actuators seemed all right. She was not moving, but looked as if she was simply waiting for orders. The sudden change in her surroundings had caused her to go into standby."

"That was Lulu, wasn't it?", he asked. Then added "but how was she able to take control of that robot?".

"Yes, it was Lulu. Madame realized that the robot was staring at a cabinet close to the doorkeeper's table and thought that the robots' emergency controls might be kept there. The cabinet was broken, but its contents seemed all right. Anyway, the robots' controls were there and, after a few attempts, she found Lulu's. I don't know how, but she succeeded in getting control and came back to the school with her."

"Yes, of course", he said. "Robots go into standby in case of serious accidents, and when they receive an order they obey without checking the authorization of whoever gives the command. I didn't imagine that RGs have such functions, which are typical of robots working in construction projects. They are designed to allow rescuers to use any robot that is close to the zone of an accident. After all, it is a good idea: an RG is just a robot, and in case of need can be used for any task".

"That's right. If it were not for that function, neither of us would have survived. They were able to get me free, and we ran towards the lowest levels of the base. There were still explosions and we could hear the hiss of air that was escaping from the corridors. We went to the cellar of the biology lab and we locked ourselves in the sealed chamber, which had been built to keep dangerous biological specimens and was empty. Before sealing ourselves in, since there was still air in the lab, Madame went out and found some food, water, oxygen and treatment for my wounds."

"How long did you stay locked in there?", he asked.

"A long, long time", she answered. "The second day Madame sent Lulu to look for something useful. We could not get outside, since the base was by then fully depressurized, except for the sealed room where we were locked in, but a robot doesn't need to breathe. I remember that she said exactly that: 'something useful'. But what is useful to an RG?"

"What is useful for her job ... and perhaps spare batteries", Mike answered.

"Right: that machine went to the night club and came back pushing a trolley with four big crates. After wasting a lot of air to get them through the airlock, we realized that in three of them there was an RG, dismantled but ready to run, and in the fourth one there were a lot of useless things like tiny dresses, like this (pointing to the one she was wearing), video projectors with a set of porno videos, and other things like that. Luckily there were also plenty of spare batteries that Madame could connect to the lighting system. At least we did not have to stay in darkness"

"You could not expect a robot to understand such a generic order. And RGs are quite specialized robots, they understand only a reduced number of words …".

"Ja, we realized that quite soon. To send her to look for food, we had to tell her 'Darling, can you please prepare us a romantic dinner?' And by sending her to look for metallic, oblong objects with a white stripe we could get oxygen bottles. To cut things short", she concluded, noticing that he was again looking at his watch, "we remained locked in there for more than two months. When we had almost lost any hope of being rescued, we heard noise coming from the upper levels."

"In that you were lucky", he interrupted her. "That moon is so far from everything that it is hard to believe that a rescue crew got there in only two months".

"Yes, and actually they were not rescuers by any means. We realized it when we managed to make them understand that someone was there. As soon as four of them got into the airlock we realized that they were raiders, looking for anything of value that still existed in the wreckage".

"My God!", he exclaimed, "it's hard enough to believe that you survived a replicator, but how did you survive those bandits?". Where replicators destroy the structure of society, criminals of all kinds emerge, and try to exploit the situation.

"When they saw what there was in our crates they started laughing, and their boss ordered the others to take them to their ship. One of them aimed his gun at Madame's head and asked for permission to shoot us, but the boss told him he was an ass, and ordered him to put his gun away and to take us outside into the unpressurized lab. With all those people who died due to depressurization, nobody would think that we had been killed by raiders, so they would have been safe from a possible hunt by some overzealous policeman. The guy who was still keeping his gun aimed at Madame shouted that he would lose the chance of killing us personally, and no police would ever discover our bodies. Anyway they would have been tens of parsecs away, by then. They started shouting at one another, but in the end the boss prevailed. They were starting to open the airlock to pull us outside, when Madame tried a desperate move. 'You are making a bad mistake, boys', she said in an insinuating voice, 'why settle for a half-dismantled RG and three you do not even know how to re-assemble? Here you have a human who can satisfy your wishes in much, much better ways'."

She paused shortly, then went on "I was shocked. I had never imagined that the director of our school would say such things, but I understood she was trying to save me. They paused for a moment, and I realized that they thought that we were the madam of a brothel with one of the girls and four RGs. At the end their boss answered: "You are right, you old slut. Why should we settle for RGs when we can have a real girl?' He ordered his men to get a space suit for me and to get Madame out, insisting that one of them remained with her until he was sure she was dead. At that point I realized it was up to me to do something. I told the boss that I didn't know how to control the RGs, and that they had better keep her alive. I added also that I would work much better if they didn't kill her. At the end they agreed and brought in two space suits and we were taken to their ship."

"We remained with those criminals for almost a year, raiding what remained of scientific stations and mining outposts that had been destroyed by replicators in the planetary system of BD -03 2001, and also in other neighboring systems. We also attacked some small outposts, adding our own destruction to those caused by those alien machines. We still have a full documentation in the recordings in our RGs. They were not aware that robots record everything, and we were able to gather such a wealth of evidence that would have had them in jail for the rest of their lives—or worse."

"While our holds were filling with more or less precious goods, they were becoming bolder, and went closer to zones where replicators had not yet disrupted civilization. They needed to dock with a commercial outpost where they could sell their bounty and refuel. One day however we got a border patrol on our tail. They followed us, and when they were joined by a police corvette, they attacked us. Our captain managed to escape, but the ship was badly damaged, and only four of the twelve raiders were still in any condition to fight. We succeeded in getting to the BD 03 1061 system, losing our pursuers, and docked at a mining outpost, where the raiders tried to overwhelm the miners and take control of the outpost."

"The miners fought back and killed the raiders one after the other. When they found us, we had to convince them that we were not members of the gang and only the RGs' recordings saved us from being lynched. Anyway they decided to take the goods the bandits had seized, then they scrapped the ship and cancelled all the recordings that they could find. At that point we could tell nobody where we came from, and we were forced to hide in their outpost. Actually they never accepted us, but for the fact that they liked the services of our RGs …" From the way Ann blushed in saying that, Mike understood that they had appreciated something else as well.

"Bastards", he whispered.

"You cannot blame them very much. We were somehow connected with the raiders, who tried to kill them and seize their outpost. And then the life of those miners was so hard, with the constant danger of being attacked by replicators …".

She paused, then continued: "Anyway, thanks to the work of those RGs we were able to live, and also save some money. We gave one of them full time to a technician, who as payment bought spare parts for Lulu and made her fully operational again."

'Now it's clear why when I started working on her she was so badly out of order', Mike thought, but said nothing.

"After two years and a half we succeeded in buying a passage on a cargo ship and arrived here. It was only a little more than 3 parsecs, but we spent all what we had saved".

"No great surprise there, when you think of the cost of interstellar travel", Mike commented.

"We have been here for more than three years", Ann went on, "trying to save enough money to buy a passage to a planet where Madame could return to being the biologist Ruth Donovan and I can resume my studies and lead a normal life. Now you know everything about us", she finished.

Mike got up and hugged her. "Can you forgive me? But all the appearances …". Then he composed himself and said: "Now I am going to leave with the Hunter. When we get that replicator I will get my share of the reward and I will be rich. It's really a lot of money: I will pick you up here and we will go to somewhere where we can live in peace. And obviously we will take Madame with us …"

He had spoken on the spur of the moment: in fact, his dreams were very different. He wanted to become a hunter. His dream was to buy a ship of his own one day: Captain Mike Edwards. However, for now he didn't want to think about that. After all, the two things were not so contradictory. For the time being he had to follow his instinct, later he would have time to make up his mind.

"No, Mike. I don't want you to take such a risk. The hunt is a dangerous business and now that I have found you I don't want to lose you. Stay here with me: give the coordinates of that asteroid to the Hunter and he will be able to destroy it. You will still get a lot of money, enough for us to start a new life here or somewhere else. Leave hunting to the professionals", she pleaded.

He was uncertain: why not stay on the station? All his world was there: his work, his friends, and now Ann. But he knew he had to go. He had to destroy that replicator, and now he had yet another reason. It was a replicator that had ruined Ann's life. He had to go, for himself, for Ann and for all those who had suffered because of those bloody machines. "Ann, the danger is not restricted to hunters. Moreover, when a hunter meets one of those machines, he is ready and has all the weapons he needs."

"The ones who are in real danger are people like you, civilians". He realized that he had already made his choice: now he was a hunter, the others were civilians. "You are well aware of that. I am the one who is afraid for you: if we don't get rid of that replicator before it generates a new one, nobody will be safe in this system any more".

He got up, picked up his bag and took out of it a small metal box with two trailing wires. He gave it to Ann, who was looking puzzled. "This morning I spoke to Joe and he promised me that he will warn you if the long-range sensors detect a replicator. If you receive a message without any content, only Joe's signature, connect this interface to your communication screen: If it shows you a message saying 'your guest will arrive at ... it will mean that, at the stated time minus 8 h the replicator will be here. Remember to subtract 8 h. At that point you and Madame go immediately to Steve's. The agreement is that they, you and Madame will board an emergency capsule and will leave this space station. We did some work on that capsule so that it will be almost invisible to the replicator's sensors. There is enough room in the capsule for all four of your, plus as many objects as you want, so you can take the RGs with you. I promised them that you will allow them to use the RGs, just to add some motivation", he concluded smiling.

"I don't think they will take us with them. You know full well that they cannot bear us", Ann answered.

"No, they've promised. And then I've left them all my savings, with the agreement that they will take you to a safe place. And it's not a small amount of money: this year I spent almost nothing and the wages of a second-class technician are not that bad. Have no fear, if the station is in a danger, they will get you out".

Ann was moved. "You should not have given away all your savings for us. And I thought that you could not stand Madame ...".

"That's true", he replied. "But I was sure that you would never leave without her. To save you I had to save her too." Then he added, in a whisper, "I love you, Ann".

"And I you", she answered. "But before you go, you still have to tell me one thing. What did you do with Lulu, while you were working on her?", she asked.

"Are you jealous?", he wondered.

"Terribly", Ann had to admit.

"But how can you be jealous of a robot? Lulu does not exist. The only thing that exists is a robot of the RG 46/G type. It's not a person, it's a machine."

"Well, you must admit that it's a machine that plays its role very convincingly. And you spent a lot of time with her. Mike, what did you do with Lulu?".

He didn't know whether to be sorry or amused by the turn the conversation was taking. How could anyone be jealous of a machine? "Nothing, absolutely nothing. Sure, I have seen her performing a striptease a hundred times, or even more, I have seen her making the same movements, a hundred times, each time in a slightly different way, while I was adjusting her actuators. You cannot imagine how such things can become boring. And then, if you still have any doubts, ask Madame to download her memory and you will see that I did nothing with her."

"What do you think? Every time you brought her back I had her memory downloaded by Madame. But that is not the point. If you went over all her programming, you could well have erased those parts of her memory that contained anything compromising. You cannot deny you are able to do that!"

It was useless to deny it. "I think so, I could do it. But, believe me, I didn't. And you must trust me …", he concluded. Then added: "But if you want me to go on working on her, sooner or later I will need to test her. Believe me, the actuators of those RGs are quite powerful. If I don't test her, she could get you into trouble. Think what could happen if she makes a wrong movement and breaks a customer's bone …".

"No, Mike, please don't", and then she paused. Then she smiled, as if she had had a great idea. "Have your friends test her … I am sure that's a risk they will be eager to run. I don't believe they will refuse!"

He laughed. "Actually I am certain they would take the risk … Nice idea, if that makes you happy, I will do it that way".

She got up, went to the door and, putting her finger on the scanner ordered: "Lock the door". Then she turned toward him. "And now, Mike, I have to show you that you were wrong …"

He was confused, not understanding what she meant with these words. Or rather, he couldn't believe that what he thought she meant was correct. "When was I wrong?", he asked.

"When you said that Lulu was much better than myself", she answered, slowly opening her dress.

6

When, three quarters of an hour later, he left Ann's room, Mike was in a daze.

Walking like a zombie, he started toward the docking area, without thinking of anything. Then he took a look at his watch: he still had ten minutes left, enough to say a few words to Steve. Now it was absolutely essential to talk to him.

He reached the canteen and knocked hard on the door of the kitchen. As soon as Steve came out, he started: "We were wrong about Madame, and you were even more wrong about Ann".

Without giving him time to say anything, he told his friend the whole story. "And you believe her?" Steve replied as soon as he could say a word, with a disbelieving expression.

"Steve, if you only could listen to her ... Obviously I believe her. Imagine what they had to suffer. And all because of those bloody replicators. You must promise me that you will help them get out of here, if anything happens." "Mike, we have already promised you that. Even if I am not really convinced that what she told you is the truth."

"Thanks, I will be back as soon as that replicator is history", he answered, running away.

The loading operation had already ended before he got on board. He rushed through the docking passage and, still running, he reached the bridge.

"You are late, boy", the Hunter told him, pointing him to a seat.

Mike tried to find an excuse, but a stare from the captain made him say nothing. He sat down and started looking around. He had never been on the bridge of a ship before, apart from his family's small cargo vessel, and this was completely different. He could see the navigation desk and the communication console, but the other consoles were new to him.

He put his bag under the desk in front of him and switched on the force field holding his body against his seat. He had learned those movements even as a child, and went through the whole procedure automatically.

"First mate, take us out to space", the Hunter ordered.

"Undock the passageway and release the docking cables", Romero said immediately.

Mike carried on looking around. The first thing he noticed was that only six out of the seven people who were supposed to be on the bridge were there: the weapon console was empty. So it was true that Thor Beaumont was so reluctant to obey the rules that he came to the bridge only when he was strictly needed, and that the Hunter suffered this only because his skill with weapons made him so precious that they could not do without him.

Mike had nothing to do, so he went on observing the people around him. He was no longer stressed as he had been at Steve's, and was quite at ease watching them. It was clear that this crew was from the inner areas of the colonized zone. You could tell that from how they dressed and from the general look of the people, but also from its composition: the Hunter made a point of showing that he had no gender bias, and three out of his seven officers were women. In the outer areas, where Mike had always lived, he had never seen any such arrangement.

He could now see the people that he had seen so many times in pictures. He knew all their names, and it was only now that he realized how nowadays names had no relationship to people's physical appearance. After centuries of cross-breeding, the features of the various human races were so weak that you couldn't recognize them, while new features were emerging because of the extraterrestrial environments where human beings were now living. He raised his eyes, and his attention was drawn to a line of large golden letters above the screens that were showing space around the ship: ... *controlling and destroying interstellar von Neumann machines is then something to which every civilization ... would be likely to devote some attention.* Below there was the signature: *Carl Sagan*.

He knew that sentence by heart and every time he thought about it he was amazed to think that scientist had understood this danger three hundred years earlier, at a time when replicators, or Von Neumann machines, were just a theoretical speculation.

He lowered his eyes and looked at the screens. The space station, which had been his home for a year, was receding, at first slowly and then faster and faster. 'Ann is there', he thought sadly. Then he cheered up, thinking that it would be just a short journey. 'We will blow up that bloody replicator and I will be back ... After all, I am also doing this for her', he concluded, thinking to all the things he could do with the reward money.

The Hunter's voice put an end to his thoughts: "Set course to the coordinates the boy gave us. Full speed".

The first mate gave the orders and the starship started gaining speed. However, nobody could feel the acceleration, owing to the inertial compensators, and the captain got up from his seat.

"Boy, put your gear in your quarters. In a quarter of an hour, report to the wardroom", he said to Mike.

"Yes, Sir", he answered in a subdued voice, but the Hunter had already left.

7

Exactly fifteen minutes later, Mike knocked at the wardroom door, which opened immediately.

The Hunter was alone, sitting at a desk. He gestured for Mike to sit down, looking at him with an quizical expression.

"I saw how you got ready for the ship to leave the space station, and I dare to say that you are quite used to spaceflight", he started on an inquisitive note.

"Yes, Sir", Mike answered. He realised that he had been summoned to be questioned. And he should have expected it. He had been taken on that ship straight away, because he knew the coordinates of the asteroid, but clearly the Hunter wanted to know as much as possible about him. "My father owned a small cargo ship, a small family enterprise. I lived on that ship practically ever since I was born."

"Well", the hunter interrupted him. "And what were you trained to do?".

"My brothers and myself did all sorts of work on board. The more we could do by ourselves, the less we needed to hire paid workers. You know, Captain, that the margins of these small transportation companies are so small … I learned to be a navigator and to perform the simplest maintenance operations. My father was very proud of me, Sir", he concluded.

"Well, here we don't need a navigator … Our navigator is quite good … But, as I told you, an extra hand for maintenance can be useful".

"I understand, Sir. I know that Ms. Bhagwat is the best navigator of all the hunting ships, and I will be honored to work under Mr. O'Connor", Mike answered.

"It looks as if you know all my crew by name. What else do you know about us?"

"Sir, I have read all I could find about you and your ship. I have seen the serial *The Hunter of Replicators* several times, even though I must say that I did not find it very realistic, being devoted to the spectacular aspects of the hunt rather than to actual facts"

"I see. You are one of those boys who dream of adventure and have romantic ideas about the hunt. I was wrong to allow you on board …"

"No, Sir. Put me on trial. I have no romantic ideas. I know full well what it is about and what are the risks that it involves. But I had to come with you. I want to be a hunter, Sir", Mike answered, clearly showing all his determination.

"But why?", the Hunter asked him.

"Because I hate those machines", he answered in a low voice.

"Someone you knew was killed by a replicator?", the Hunter asked again.

"When I was eighteen I left our ship to go to the University on 82 Eridani IV. My father was saying that I should get a degree so that I would have a chance to live a better life than you can have on a small cargo vessel—and above all, I guess, because he feared that it was no longer possible to get a living from interstellar commerce in a society that, due to the increasing presence of replicators, is falling apart."

The Hunter interrupted him: "Well, you were not so badly off if you could afford to go to University on 82 Eridani IV, just 6 parsecs from Earth".

"No, Sir, we could never have afforded it if I had not got a scholarship", Mike answered, and continued: "six months after I left, my ship met a replicator and was destroyed. There were no survivors. Now you know why I hate them".

"You cannot hate a machine. Remember that they are just machines, extremely dangerous ones, machines we must destroy, but nevertheless just machines", the Hunter interrupted him again. The boy's story had impressed him, but he felt that that kind of attitude could be dangerous.

Mike was impressed by his words, too. They sounded similar to what he had said to Ann, when speaking about Lulu: "You cannot be jealous of a machine, RGs are just machines".

Humans spent all their lives surrounded by machines, some of which behaved like humans, but nevertheless were just machines. Machines that spared humans heavy work, like hundreds of different types of robots; machines that gave them pleasure, like RGs; and machines that kill, like replicators. Machines you used or you had to destroy, but that cannot be objects of hate, love, jealousy ...

"I should have left you on the station" the Hunter started to say. "If you hate replicators you will never be a hunter. Or at least, you would not remain alive as a hunter for very long. When you are confronted with them you must have a clear mind—sooner or later hate will lead you to make a mistake, and you cannot afford mistakes. You must destroy those machines in just the same way you must stop a boulder that is falling on you. No hate, only the consciousness that it is your duty".

"Yes Sir, I'll remember that", Mike answered. "But many say replicators are not just machines, but that they are intelligent beings: they are conscious and have their own will", he continued, knowing full well that the Hunter had always opposed these ideas.

"What do you know about replicators?", the Hunter asked him, without entering into any discussion of that point.

"That they are intelligent, self-replicating machines, able to move from star system to star system. They are the interstellar Von Neuman machines Carl Sagan spoke about centuries ago", Mike went on, remembering the sentence written in golden letters that he had seen on the bridge.

"We only know they were not built by humans, and thus they must have an alien origin, which is strange, because we have never met intelligent beings of any kind. The only extraterrestrial lifeforms we have found are bacteria, and quite primitive bacteria at that. Nobody knows who built the first replicator, perhaps millions of years ago. From all appearances, it might have been a civilization that flourished in a distant past in the Perseus arm of the Milky Way, in the general direction of the constellation of Canis Major. One of them enters a star system, lands on an asteroid that contains the required raw materials and starts replicating. Its offspring explore the system and in turn replicate until one of them leaves, heading towards a nearby system".

Mike paused shortly, and then continued: "If during their exploration they find organic life forms, and above all, higher life forms, they destroy them. The first encounter took place in a frontier system in the Eridanus constellation, and they are moving—luckily slowly, because they cannot travel faster than light—towards the more densely populated inner zones. If they go on like this, they will reach the Solar System within a hundred years." With that, Mike had said practically all that he knew.

"Actually you have condensed in a few words all that we know about them. I would really like to know who did set these machines—which are creating havoc—free to roam our galaxy. Do you know that in Sagan's time there was a discussion about Von Neuman machines? Another physicist, named Tipler, held that interstellar Von Neuman machines were the best way to explore the Universe. He even said that a civilization able to build such machines could explore, colonize and even control the whole Universe".

"That's madness. It might be able to wreck the whole Universe. Luckily there are people like you who try to stop them", Mike countered.

"There should be a better understanding of the danger. Governments do practically nothing, except for contributing to the fund for rewarding anybody who destroys one of them. But perhaps it is not the wrong strategy: it costs much less to pay rewards than to build a fleet to deal with the problem", the Hunter said.

"But is it true that the money from the rewards is enough to keep a ship like this and her crew in operation?", Mike asked.

"Yes, certainly. And if you manage to avoid heavy damage during the hunt, you can make a lot of money", the Hunter added. "But now let's deal with our replicator. How reliable are the coordinates you gave us?", he asked.

"I don't know, Sir. I am sure of the coordinates of the asteroid at the moment it encountered that miner, but ephemerides in this system are quite imprecise. And in particular, the asteroid was located in the zone where the gravitational attractions of the two stars are in equilibrium and any orbits are quite unstable. I reconstructed the orbit using the best ephemerides I could find, but the errors could be quite large …", Mike answered.

"Let's hope they are not. If the position is correct, we should get there just six hours before the replication process is finished. And if we need to waste time looking for it, we are in trouble".

"If time is so critical, I could make some small alterations to the engines that would allow us to save a few hours …"

The Hunter stopped him. "Or blow us up trying to do so. No chance, boy. How do you think you could make this ship faster?".

"I've looked at the make and type of the engines and I discovered that they are identical to those of the ship of a bootlegger who docked at our station some time ago. I copied all the schematics of those engines and I can make all the modifications they had made to increase the speed", Mike said quietly.

"And you worked on an outlaw's ship? Don't you know that I could have you arrested for that?"

"Yes, Sir, but I did it only because the head of maintenance ordered me to do so … And I've told you because this is a way we can save some time …", he answered, becoming more and more worried.

"And how much did you get for that? The same as when you tried to sell those RGs and that … human to my men?", the Hunter asked with obvious contempt.

"Nothing in both cases. With the bootlegger's ship, the maintenance chief got all the money. As for the RGs, I did it as a favor to a friend. And about that girl, don't think badly of her. She is not a whore, she and that older woman are the only survivors of an attack by a replicator, and if they live by renting out those RGs it is only because they lost everything and are trying to survive".

"Never mind the girl, who is undoubtedly a bad lot, to go around dressed like that. Why did you work on that ship, if it was not for a bribe?".

"Because my boss ordered me to, Sir. I couldn't refuse". Mike could not understand why the other was making such a fuss about repairing a bootlegger's ship.

"Surely you could have done. Remember that you must not obey unlawful orders, whoever gives them to you. Even if it's me, I mean."

Mike looked at him in amazement. How could he fail to obey his captain? Then he remembered that the Hunter had to leave Starfleet because he refused to obey an order he considered unlawful. He had better get out of that minefield pretty quickly. "You will never give me unlawful orders, Sir", he answered. He didn't like the turn the conversation was taking at all. "I hope not", the captain replied. "Now go to the engine room and see whether you and Mr. O'Connor can soup up this ship."

Before leaving, Mike turned towards the wall above the door, where six plaques were displayed. They had been awarded for the six replicators the Hunter had destroyed.

"Within a few days there will be a seventh plate on that wall. And you will be awarded the *Order of Sagan*", Mike said smiling.

"Remember that a replicator is not destroyed until it is there in space, in front of you, in pieces so small that it is not worth even selling them as scrap metal", the Hunter told him dryly. "and that saying such things brings bad luck", he added in a low voice.

"Good hunting, Sir", Mike concluded, feeling more and more nervous.

"Good hunting, boy", the captain answered.

8

When Mike got to the engine room, O'Connor was performing the last checks before running the engines at full power to get the maximum sub-light speed. Because their whole course was inside the stellar system, they couldn't use the warp drive.

"The captain told me to come here. He ordered me to make a few adjustments to the engine to increase the top speed", he told O'Connor, handling him the memory card to which he had downloaded all the data he had acquired from the bootlegger's ship.

"Are you crazy? I won't allow you to touch that engine, even if I get a written order from the captain", O'Connor answered.

Mike told him the story of the bootlegger, but it took him more than twenty minutes to convince O'Connor to even take a look at the data. However, as the various drawings and schematics were displayed on screen, the attitude of the chief engineer gradually shifted from hostile to skeptical.

"It looks as if we could try it ...", he said after some time. "Since every second is precious, we may as well see if we can do something", he concluded.

They worked the whole afternoon, and at six, when they tried to run the engines at full power, they realized that the ship was faster than usual.

"I'll tell the captain you did a good job", O'Connor told him while they were walking toward the mess hall for dinner.

Mike was elated: he felt he had earned the respect of the chief engineer, and hoped to be accepted by the whole crew. His dream of becoming a hunter was on its way to becoming true.

When, after listening to the chief engineer, the Hunter congratulated him and told him he had decided to reinstate him to the rank and pay of second-class technician, his spirits really soared. Not so much for the pay rise, because if they actually got the replicator, his share of the reward would still have been more than the rise, but because that meant that the Hunter was happy with his work.

At the end of dinner, Romero took him aside and said: "A nice swindle you pulled on me with that RG. It was true about her voice, you made an excellent job there: not even the RGs you find in the brothels on Earth have a voice like that. And also the striptease was not at all bad. But when you get to the real thing, she is really mean. She moves like a ... a ..." and he stopped, not finding any comparison that could adequately express his dissatisfaction. "When we get back to the station you will never be able to convince me again to go with her: I want that human, no matter how much it costs. Look: with all the sweat and the danger we have to get through to earn the small amount of reward money, I will not settle for a bunch of actuators worked by a computer and covered with synthetic skin." Initially, he seemed to be joking, but now he was getting furious.

Mike tried to hide his satisfaction. "I told you, I wasn't able to finish my job. I have still to adjust the actuators that drive the pelvis and the heaters … as you know full well. Let me finish my work and then you will see …" Then he added: "As for the girl, I have bad news for you. She has quit doing that job". He refrained from saying 'actually she never was what you think', because he didn't want him to get too angry.

He immediately went to bed but couldn't sleep for hours. He was too excited: all his dreams had suddenly become true—all of them. He was on his way to becoming a hunter; he had been accepted on the best known hunting ship; he was on the trail of a replicator; and soon he would be rich. Ann was not what he had feared; and she loved him. How could he sleep with all those thoughts on his mind?

9

Over the following days they fell into the usual spaceflight routine. The replicator was still a long way away and there was not much work to do. The Hunter remained in his quarters for almost the whole time, and everybody was trying to pass the time however they could. The first day Mike was really busy with the maintenance operations that were usually done at the space station, but soon that work had also been finished.

Only then did Mike realize that travelling on that sort of spaceship was completely different from what he had known during his childhood. On their small cargo ship they had always had something to do, and the watches spent on the bridge were a long and tiring business

The ship, an old, but perfectly refurbished Starfleet destroyer, was on the contrary, in perfect condition. Everything was clean and tidy and the crew was large enough to allow a second-class technician to just give orders, without having to do much work personally.

Mike was getting bored and, to do something, downloaded to his computer the robotics textbooks he had used in the few months he spent at the university on 82 Eridani IV. He realized that, now that he had first-hand experience of fixing or re-programming robots of various kinds, theoretical knowledge started to take on a different meaning. And a better knowledge of robotics might perhaps allow him to understand replicators better.

He became friends with Takashi Campos, the scientific officer, and spent many hours in his lab discussing the nature of replicators, but soon realized that the scientist didn't know much more than himself on the subject. And that was notwithstanding that he had spent several years studying them. In practical terms, all that was known about replicators was what he had himself said to the Hunter in the wardroom.

The eighth day after they left the station everyone started to become nervous and started to feel the excitement of the hunt. Well before 4 pm, the predicted time of the rendezvous with the asteroid, everyone was on the bridge. Everybody except the Hunter, obviously.

This time Beaumont also was there. Sitting at his station, he was running final tests on the weapons systems.

At half-past-three the alarm rang as expected, and immediately after that the Captain entered the bridge, and sat down at the command station. Straight away he ordered "Passive sensors at maximum range and active sensors off. The replicator must not detect us before we determine its position". The procedure was always the same, but the Hunter liked to give orders as if it were their first hunt. Perhaps sticking strictly to the rules was simply one of the superstitions that no hunter was able to suppress.

"Passive sensors at maximum sensitivity. Nothing recorded", Campos answered.

They carried on for half an hour without speaking, until they reached the point where the asteroid should have been located.

"Full stop. Start probing the surroundings with active sensors, at increasing power. In case of a contact, switch the sensors off".

Mike had read reports from hunters several times, and knew that was the standard procedure. It would take another half an hour to probe the surroundings of the point that they had reached.

Time seemed to drag. After half an hour Campos announced: "No object within the range of active sensors". Then he commented "There is nothing here: not an asteroid, not a ship, nothing."

"It looks like your coordinates were wrong, boy", the Hunter said addressing Mike. Then, seeing his disappointed expression, he added "Don't worry. I never managed to find one of those machines at the first attempt."

They plotted a set of equal-probability spheres about the point they had reached and started moving slowly toward the region where the asteroid was most likely to be found, with active sensors at maximum range. After more than one hour, Campos said in a low voice "I'm getting a very weak echo, very, very weak. It doesn't look like an asteroid … It might be a metal object".

"Let's close in cautiously. Reduce the power of the sensors, weapons ready and shields at minimum power, to reduce our radar cross section", the captain ordered.

It was probably a ship, perhaps a mining ship, even if it was strange that miners were still there, after a replicator had been spotted in the area. Or perhaps it was just a wreck: a miner less lucky than the one who had been taken to New Shanghai. But there was always the possibility that the replicator had finished building a copy of itself, and then it could be one of the two machines.

Now the tension on the bridge was palpable. Nobody said a word and all stared at the screens.

Twenty minutes later, the telescope finally showed an image on the screen: a spaceship not very different from the one they were in. It was certainly not a cargo ship.

"It's probably a hunter …", Romero said.

'No, let's hope not', Mike thought. He didn't like the idea that there was competition, and the possibility that they might have to share the reward.

"We'll play it carefully, it might be a bootlegger, or a raider. Close in carefully. Sensors and shields at maximum power", the Hunter said.

"It seems to be damaged", Campos noticed as soon as the image improved. "I'd say it's a wreck".

"Ms. Chen, try to get in contact to identify that ship", the Hunter ordered.

Liza Chen started sending out identification requests.

After a few minutes a voice came from the loudspeaker. The transmission was quite garbled and there was no video.

"I am Father Dubois. We were attacked by a replicator. We request assistance".

"No, that heretic is exactly what we need now!", Romero howled. "Let him go to hell, he and all his crew. Let's carry on with our search".

"We cannot leave them in space, on a ship in that state", the Hunter replied.

"But Sir, if we give them assistance, we will never reach that replicator before …", Beaumont said.

The Hunter stopped him. "I said we cannot abandon them in space on such a wreck". Then he gave his orders: "Maximum speed on an intercept course, but carry on probing space in all directions. Give me a communications link".

He picked up a microphone from his console, and started speaking as soon as Chen told him he was on the air: "Father Dubois, get ready to be boarded. Do not try to fight back: we will take you on board. I promise I will release your crew as soon as we reach a space station, but I will deliver you to the authorities on the closest planet".

"Do not worry, you know full well that we have no weapons on this ship. We cannot do anything but surrender to you", the voice from the loudspeaker said.

"Get a boarding party ready. When they are on board here, they must be confined. Father Dubois must be put in isolation, but treat him well". Then, knowing how much his people hated Father Dubois, he added "Treat them well, it that understood?' Everybody understand?"

Romero took a microphone from his console and ordered six crew to go to the ordnance depot. Then he said to Mike "Boy, can you use a gun?".

"Yes Sir … at least a little", Mike lied. Actually he had used a gun twice at a shooting gallery in an amusement park, as he said to himself in justification.

"Go with them", the first mate ordered, getting up and going to the armory himself.

Mike joined the others, got a gun and put on a space suit, then they all trooped to the airlock.

After a few minutes they heard a thud, followed by the noise made by the connection tube that stretched out to link the two ships' airlocks. A few minutes later, a green light came on over the outer door: the pressure on the two sides had been equalized.

"Seal your suits and get ready for boarding. Keep your weapons at the ready", Romero ordered.

As soon as the door opened they rushed into the corridor and in a few seconds they reached the other ship. In the airlock there were eight people, all in fairly bad condition. Many had bags with them, clearly containing all their belongings, but nobody had weapons.

With their guns leveled, they surrounded the crew of the disabled ship. One on each side, two of them went to cover the door leading to the interior of the ship. Mike took a closer look at the crew and realized that they really were in a bad way. A man and a woman were sitting on the floor and had bandages on their legs. All of the crew had bandages, and many were still stained with blood. An older man, in a black suit with a small white metal cross pinned on the collar of his jacket, and who had an arm in a sling, turned to Romero and said quietly "You can put away all those weapons: we are no danger to you. As I said, we have no weapons on board".

"Where are the others?", the first mate asked without answering.

"Dead. We buried them in space. We are the only survivors", the father answered.

Romero was not so easily satisfied. He pointed to four of the men from the boarding party "You check the ship. And you, boy, go to the bridge and download all the data from their computer". Then he gave his orders to their prisoners:

"Move now. We will check you one by one and then we will lead you to your cells.".

Mike moved quickly toward the bridge, but at a certain point he found his way blocked by a hatch that was locked tight. The red light above the door clearly indicated that the zone on the other side was depressurized.

At first he thought of blowing the door and to go on by using his space suit, but then he realized that he didn't need to. He went back until he found a compartment where the door was open, and went in. The room was a mess, and the bedclothes were in a tangle: clearly the person who lived there was asleep when the replicator attacked the ship and he had abandoned his room in a hurry.

He went to the computer terminal and tried to switch it on. It still worked. In less than a quarter of an hour he was able to get access to the mainframe and to download all the navigation data and the ship's log onto a small memory card he had with him. Father Dubois had not thought of putting any specific protection on the memory areas where the navigation data were stored.

He put the memory card with all the data he had acquired into the backpack of his suit and went back. The airlock was empty: they had all gone back to the ship. As soon as he had passed the interconnecting tunnel, he closed the outer door of the airlock and heard the noise of the docking clamps opening: Father Dubois' ship had been abandoned in space.

He thought it was a pity: the wreck, although in a pretty bad condition, was still worth a lot of money. However he realized that they couldn't tow it to a space station if they wanted to continue hunting the replicator. He went to the bridge and as soon as he got there he realized that everyone was at their station and that the ship was again scanning the surrounding space.

"Sir, I got all the data from that ship's log", he immediately told the Hunter. He got out the memory card and gave it to Bhagwat, who inserted it into her terminal.

"Well done, boy", the Hunter told him. Then he ordered: "Ms. Bhagwat, study those data and try to compute the course the replicator that attacked the Father's ship took after the kill. Mr. Campos, download the images of the battle and try to make sense of all that you find out."

For a quarter of an hour nobody spoke. Then the image of a replicator came on screen. "Here it is, Sir", Campos began, "These are the images of the battle—if we can call it a battle. They were taken thirty-six hours ago".

The sequence was short: the replicator started shooting its plasma jets from a great distance, when the ship was still well beyond its maximum range. After many misses, a plasma jet finally struck father Dubois' ship, after it had started accelerating to get out of range. The hit was not as severe as it might have been and didn't affect the engines, so that the ship was able to go on gaining speed. A second hit produced more damage, but the ship managed to escape.

At that point the replicator gave up the chase and returned to its previous course. It was at that moment that an explosion wrecked the engine room, and subsequent explosions, triggered by the first one, damaged other parts of the ship, which was reduced to a wreck. It continued on the same course while the replicator headed in another direction.

"It's strange behavior", Romero said. "It started shooting too early, wasting a lot of plasma. And, above all, its aim was bad. And then it gave up the chase too early. If it had hit the engines with its first shots, it would definitely have destroyed its target."

They remained silent for a few seconds, then the Hunter said: "It is clearly an inexperienced replicator. There is just one explanation: it is the offspring of the replicator we are looking for. The replication process ended earlier than expected and the new machine left immediately. Then it encountered Father Dubois' ship before it could test the command sequences in detail and before it finished its self-training sequence".

"That bastard always has luck on his side", Romero commented. "I have bad news", Bhagwat announced. "From the data on its course, it is clear that after the battle the replicator set course toward the inner parts of the system, in particular toward the star's B component".

"Are there inhabited zones in that direction?" the Hunter asked.

"Yes, Sir. It looks as if it is aiming toward the third planet, a gas giant that has a satellite on which there is the Ceres agricultural colony. There are almost thirty thousand people there".

"Set course towards the third planet, then. Our original replicator can wait: usually after a replication cycle they need about sixty days to carry out all the tests and refits before they move. We have time to get the new replicator and then we can go back to deal with the first one."

"That's great: two rewards in a single campaign!", Romero exclaimed.

"Quiet, Mr. Romero. We must succeed in getting them first" the Hunter rebuked him. "Can we get it before it reaches that agricultural colony?, he asked Bhagwat. "It has a lead of about thirty six hours". "If we hurry, we should. But we have quite a small margin", she answered.

"Well, let's maintain top speed then. We must save those thirty thousand colonists", the Hunter stated.

"And if we fail, after thirty thousand casualties the government will raise the reward", Romero added.

The Hunter said nothing, but threw him a glance of deep disapproval. Shrugging his shoulders, Romero focused his attention on the console in front of him.

"Well, boy, it looks like that the adjustments you made to the engines will still prove useful", the Hunter told Mike. "And now take something to eat to our prisoners. You are used to playing the waiter, isn't that right?".

Pretending not to notice Mike's disappointed reaction to his words, he added: "speak to the Father and see if you can find out anything. A formal interrogation would be useless, and anyway he will be questioned when we deliver him to the police, but perhaps he will tell you something."

"Thank you, Sir. I will try to talk to him", Mike replied, getting up.

10

Mike went to the mess hall and had the food synthesizer prepare eight meals. He put them on a trolley and started walking toward the cells.

The prisoners were in three cells: the crew were in the first two and Father Dubois was alone in the last one.

After taking the food to the others, Mike gestured to the man who was guarding the third cell to let him in and to wait outside.

As soon as he was inside, the Father, who was lying on his bed, got up and sat on a chair.

"Father, I've brought some food", Mike said.

"Thank you, it's very kind of you to care for a prisoner like myself. And no doubt a difficult prisoner", he said. Then added "What's your name?"

"Mike", he answered. "The Hunter ordered me to bring you food. We have nothing personal against you, but you are on the list of people wanted by the police of every inhabited planet".

"I know, the Hunter is a good person", he answered, "even if he is on the wrong side. You are all murderers, but you act in good faith … God will forgive you", Father Dubois stated.

Mike put the tray with the food on the table and sat on the other chair.

"Father, eat the food before it gets cold", he said.

"I am not hungry, Mike. You know, I am very tired. Tired and depressed. I have failed, and my failure caused the death of a lot of people", he said slowly, "and I fear that others will die in the future. But there must be a way … I've prayed so much, God will listen …"

"A way to do what, Father?", Mike asked. The Hunter had ordered him to talk to the prisoner, but the Father's words were intriguing him.

"To speak to them. To make them understand that they must not kill us. Only peace must reign between the creatures of God. We must stop killing each other: after all we are brothers. We must both understand that we are brothers." Mike was amazed by these words, and above all by the tone with which he had pronounced them. He had heard about Father Dubois, even if he had never believed the stories. Nearly everyone said that he was crazy, or a fanatic, but he was not speaking like a mad-man. What amazed him was the contrast between the absurdity of those words and the simple, matter-of-fact tone that he used.

"But they are not God's creatures, they are machines. They are only machines, robots built by goodness knows who with the aim of destroying us. And we do not kill them. We destroy them, in the same way that we have to destroy a dangerous object", Mike answered.

"You are wrong Mike. You are wrong in the same way that the Hunter, who murders in good faith, is wrong, Mike. As my Church was wrong when they declared me to be a heretic. As the Pope was wrong when he excommunicated me", the Father said in a sad voice. "But sooner or later you will all change your minds. You will realize that behaving as you do has no meaning, it can only lead to more deaths. More deaths on both sides."

When Mike said nothing, the Father went on. "You see, people say they are only machines, just as you say that our robots are only machines. But they are complex machines, complex to the point that they are intelligent. And intelligence is accompanied by self consciousness, and then by a soul. When the Hunter boasts he has destroyed six replicators, he should say that he had murdered six intelligent, self-aware beings, beings with a soul like you and me. He does not realize—you don't realize—the fact, but then it is likely that God, in his infinite compassion, will forgive you. But that is not the point: you may be not completely responsible for your crimes, but crimes they are."

Suddenly Mike remembered why the Father's ideas had been declared heretical: he held that robots, when they were complex enough, had an immortal soul.

"But father, I know robots well. I have fixed many of them and I have even reprogrammed some", and here he thought to RG46/G. "I can guarantee that they are just machines controlled by a microprocessor—by many microprocessors, actually. It's the program running on the microprocessors that controls them. They act thanks to reflexes, complex as they may be, but only reflexes: if such-and-such happens, then do this. They react in a complex and intricate way, but there is no trace of consciousness, of a soul in them."

"You are wrong, Mike. How can you say they have no self-consciousness? If you could dismantle my brain, you would find only neurons, neurons without end. Naturally my brain is different from the microprocessors in our robots, or from the computer built by aliens—and nobody knows how it works—that controls a replicator. My brain, your brain, all human brains are just biological computers on which a particular software is running, exactly as it occurs with robots. But that software, owing to its complexity, is able to make us self-aware; produces our mind. And our soul is directly linked to our mind. And the same applies to them. Our soul, their souls, come from God. They are a spark of that infinite consciousness that permeates the Universe and which we call God".

"But, father, that is pantheism", Mike exclaimed. He had studied some philosophy when he prepared for the exams before entering university, and which he took by remote learning from his family's cargo ship. In fact, his best marks were for philosophy. The name of Spinoza came to mind, but he refrained trying to impress the Father with such recollections from high school, because he was too uncertain of his facts.

"Yes, pantheism. They charged me with pantheism as well, when they said my ideas were heresy", the Father said sadly. "But what can I do if this is the truth?"

"Father, I am too ignorant in theology, I am just a maintenance technician. I studied pantheism in high school, in the philosophy classes, but I cannot say I understood it. But I do know robots ..." He paused since he was ashamed to speak of the RGs. However, he knew that the Father would understand and went on. "I re-programmed a RG. You know what RGs are, don't you, Father?"

"Sure, those poor girls, They are treated as slaves, used by men for their pleasure, and sometimes broken into pieces by wicked men ... I have often prayed for them".

"But Father, they are not girls, they are machines. Machines, like ... like the control unit of the synthesizer that prepared your meal—which, by the way, is getting cold. And then, if I remember correctly, your Church has a more flexible attitude to RGs. Many of your brothers say that after all, in these remote stretches of the colonized zone, where miners and astronauts cannot bring their families, it is better that the job is performed by robots rather than by human girls".

"I know, Mike, but they are wrong. RGs, like their male counterparts that are used in the inner parts of the colonized zone where women are as numerous as men, like all complex robots, and thus like all replicators, are sentient beings. You studied philosophy: do you remember the categorical imperative?".

"That's Kant, isn't it?", Mike tried to say. Vague reminiscences from school were coming back to his memory. He didn't remember what Kant had said, but he was quite sure that he was linked with the categorical imperative.

"You deserve full marks for philosophy. Yes, it was Kant. Human beings must be an end of your actions, never a means. Human beings—self-aware beings in general—including RGs and replicators, obviously. You cannot use a self-aware being for your pleasure, Mike."

Mike remembered what the Hunter had said to him. He was there, not to discuss RGs or moral philosophy. He had to bring the subject of their conversation back to replicators.

"But, Father, replicators destroy any lifeform they meet. Even if they are self-aware, if they have a soul, it is still our right to destroy them. It is self defense, after all", Mike said.

"No, we must not answer violence with violence, murder with murder. We must bring peace to this galaxy. Just as we do not realize that they are sentient beings, they don't realize that we are. We must enter into contact with them, talk to them, convince them. Only a peaceful response can stop this spiral of violence", the Father answered.

"But nobody has ever succeeded in getting in contact with them …" Mike started.

"I know, I know", Father Dubois interrupted him. "This is why I am here. My ship was full of communication systems and as soon as we saw the replicator we started broadcasting on all frequencies."

Mike was about to comment that the results had not been very encouraging, but refrained. Instead he said: "It is not a problem of frequencies, or of other technical details. For years we have known which frequencies replicators use for the few messages they sometimes exchange between themselves. And that notwithstanding, no contact has never been established".

"So, what then? Only with patience can we get results. Peace requires patience …"

Mike wanted to tell him that replicators were expanding in the inhabited zones, with all the ensuing tragedies. They had no time to be patient. Instead he changed the subject. "Father, there are those who say that replicators come from the devil, to the point that I have heard that there are satanists who worship them as manifestations of the powers of evil".

"Our worlds are full of crazy and stupid people", the father interrupted him. "Don't tell me you share those idiotic beliefs".

"Certainly not, but I understand that there is a general agreement between religions in assessing, although to different degrees, that replicators are a form of evil. Where they arrive, they disrupt civil society and with that order and civilization vanish. Many religious institutions, like the Catholic Church, many Protestant Churches, the Muslim Caliphates, together with many states and political organizations of various planetary systems participate in funding the rewards. There are imams who have declared a jihad against replicators. And the Holy Father, when he received the Hunter and his crew, encouraged him to carry on with his battle against what he defined as 'the powers of evil that endanger humankind'", Mike went on.

"I know, I know. My Church—they excommunicated me, but I continue to consider it my Church—is wrong. The Holy Father is wrong. But they will understand. Only peace will put an end to these massacres."

Mike realized that the Father was firm in his beliefs. After all, neither a trial for heresy, nor the expulsion from the Jesuit order had shaken his ideas. He left the tray of food on the table and left, walking back toward the bridge.

As the Hunter was not there, he looked for him in the wardroom and found him studying some maps of the system.

"Did you speak to him? What did he tell you?", the Hunter asked him.

"Yes, sir. He gave me a long lecture on our need to talk to replicators to find a way of living in peace. And he says that everyone, from you to the Holy Father, from his Church to the governments and all the civil and religious organizations, are wrong".

"As usual. His conceit is unbelievable. He is sure that he is the only one to understand the situation, while we common mortals are all wrong. But in practice his solution to our problems is meaningless. What does it mean to talk with the replicators? How can you talk with a machine? And while he is trying to establish his dialogue, people go on dying. And what if he were right? All the worse: What if he were right and those were intelligent and conscious beings? They would still be an incarnation of evil and destruction. And no compromise is possible with evil. The only solution is to fight against it", the Hunter told him. His voice was plain and low, as if he were speaking more to himself than to Mike.

"But, Sir, I understand his problems with his Church … but why are the police forces of every system after him? Why do we keep him in jail and why are we going to deliver him to the authorities?"

"Don't you know that with that ship, which he was able to buy thanks to the money collected by his supporters, he has several times hampered hunters while pursuing replicators? And the true reason he is wanted is in connection with a number of sabotage attempts on hunting ships in various systems. Perhaps we should mount a closer watch on his cell … you never know."

Now Mike had a better understanding of the danger the heretic represented. The danger was not so much in his ideas, which were shared by a few people, but in the actions that a few fanatics could do. There were not many hunters around, and to stop some of them, even for a short time, meant allowing replicators to hit many targets.

"We will keep our eyes open, Sir. And next time I take him food I will try to understand whether he has anything in mind".

"Thank you, boy. But remember to treat him well. That man is wrong, but he is honest and believes in what he is doing. He deserves our respect, even if we must prevent him from putting his ideas into practice".

Mike realized that those two men, although fighting on opposite sides, admired and respected each other.

11

Little happened in the following three days. The ship was travelling at top speed towards the third planet, and everyone hoped to reach Ceres before the replicator got there. The distance between the ship and the replicator was decreasing fast, but the predicted interception point was dangerously close to the satellite.

Mike visited Father Dubois several times, but every time, their conversations became shorter: they had said all they had to say to each other. Moreover, the Father seemed increasingly depressed and withdrawn. The first day he begged Mike to ask the Hunter for permission to attempt to make contact before the machine was attacked, but the request was never repeated after Mike told him that the Hunter had refused. Father Dubois clearly understood that the Hunter couldn't lose the advantage of a surprise attack by giving away his presence too early. Like everybody on board, the father was waiting for the battle to take place, knowing that he had no role to play. He would have to witness a battle in which one of the two sides was bound to be destroyed and he realized that his personal destiny and the life of thirty thousand people were strictly linked to the Hunter's survival.

On the fourth day, the bridge started to be crowded again and an hour before the predicted contact time everyone was at their station. At last a small bright point appeared on the forward screen. "Active sensors at full power: Let's try to be detected. If it comes after us, the risk of it attacking the colony will be reduced", the Hunter ordered. From the expressions on the faces of the crew, Mike realized that few of them agreed with that strategy. "Ms. Chen, get in contact with the colony" the Hunter ordered.

Mike looked at the screen to judge the distance separating them from Ceres: they were still so far away that they would not get an answer for at least three hours. The pursuit continued without the replicator giving any sign of changing its course, although by then it must have been aware of their presence. After three and a quarter hours the image of a woman appeared on the screen. The background was typical of an agricultural colony: the communications room had large windows through which one could see a hill covered in thick vegetation. Mike realized that one of the windows was open, and that the woman had no breathing gear. The moon not only had an atmosphere—something essential for an agricultural colony, but an atmosphere that you could breathe freely. Mike was not used to a thing like that: he had lived on a planet with a breathable atmosphere for less than a year in his whole life. He had always been ill at ease when in the open without a space suit or at least breathing gear. "We are honored by your visit, Captain Hillman. If you need agricultural supplies, we will be pleased to offer you all what we have". Mike found it so strange that the Hunter should be addressed using his name, that for a moment he couldn't understand whom she was addressing. To his crew he was just the Captain, while for everyone else he was 'the Hunter'.

"Be careful", the Hunter interrupted her, "a replicator is directly aiming at you. We will try to stop it before it gets to shooting distance, but I suggest you take shelter in the safest places in your colony. I hope you have shelters safe from the weapons of replicators. Take food, water and oxygen with you and get ready to withstand a long siege before you can return to the surface". Now they were closer, and the time needed for an answer was less than an hour. In the meantime the image of the replicator had grown larger and more detailed. It was no longer just a bright dot, but was a huge spacecraft.

When he could make out the details, Mike saw that it had no true hull, but it was a sort of lattice structure with no pressurized compartments. He had seen such an image several times before, but only now did he really understand what that kind of structure meant. It was a machine built by another machine, and not designed to support living beings. It had no life-support systems, no shelter against space radiation. He had read that its shields were barely sufficient to stop the most penetrating radiation, that could otherwise damage its control systems. The image of the machine on the screen was becoming larger and larger, until it filled the whole display. Mike did, of course, know that the length of a replicator was more than 1500 m, but it was one thing to read these numbers in a book, and another to see it there, right in front of him. The communication screen came on again and the same woman who had answered their first message appeared. Now she was scared. "Are you sure, Captain? Our long-range sensors have not detected anything yet. We don't have shelters, only underground stores, but I don't think they can withstand the force of its weapons. Perhaps the best thing is for us to scatter on the surface, and find some shelter in the natural caverns that can be found in the mountains".

"Don't try to find sanctuary by dispersing on the surface. Replicators destroy all organic life, and in particular animal life: it will hunt you one by one. If you don't have shelters, natural caverns may do, provided they are deep. But take supplies with you: in the past replicators have besieged a planet for months. We will try to destroy it before it gets to you, but take all possible measures you can, just in case we fail" The Hunter closed the communication channel. "And now the hunt starts. How long will it take us to get within range?", he asked Bhagwat. Then he added "Put on your space suits and buckle up your restraint systems".

"It will be in torpedo range within fifteen minutes", Beaumont answered before the navigator could speak. By that time, everyone had donned their space suits and were secured in their seats. "Don't open fire until my order.

We must not waste torpedoes launched from a distance at which they are ineffective". Time seemed to be passing slowly now. They were all nervous and the silence could have been cut with a knife. The fifteen minutes went by without anyone opening fire. And the replicator didn't make the mistake it had made when it attacked Father Dubois's ship. 'Hell's teeth, those machines learn quickly', Mike thought. He was sure that this time its shooting would be much more precise. After ten more minutes the rear batteries of the replicator opened fire with a plasma bolt that struck their forward shields dead center. Space around them lit up, but the shields easily withstood the shock. "Torpedoes from one to ten out. Aim at the engines", the captain ordered.

As the torpedoes were completing their run, O'Connor got up. "I'm going to the engine room to be ready for any possible repairs", he said. The Hunter gestured his agreement. "Should I go too?", Mike asked. "No", the Captain answered, "stay here, ready to perform any repair that might be needed in this part of the ship". The torpedoes hit the replicator's shields without causing damage. It was clear that these were just the early phases of a battle that was bound to be long drawn out. Mike had read about so many battles of this sort and had dreamed many times of taking part in the hunt. Now that he was in midst of one he started to feel scared. The Hunter must have realized that, since he told him: "you wanted to come here at any cost, now you realize what it is all about". Mike wondered whether he actually looked so frightened. "Yes, Sir. I will do my best, Sir".

The distance was decreasing fast, but the replicator was still heading towards the moon. If they didn't destroy it within the next few minutes, it would get close enough to the surface to start bombing it with its small nuclear warheads. "Set a course to take us right behind its engines, and get the front batteries ready", the Hunter ordered. He was trying to prevent the replicator from using its propulsion system to achieve a low orbit about the moon.

When they were at close distance, he ordered open fire, aiming at the engines. The light beams from the lasers hit the replicator's shield and scattered over the surrounding space. However, when the light level was again low enough to allow them to look at the screens, they realized that the shields had not failed and the engines were still undamaged. Suddenly the replicator started rotating about an axis perpendicular to its course, and to shoot bolts of plasma whenever one of its batteries had a chance. This was what the Hunter was waiting for: it was possible to strike through the opening in the shields that it had to open to leave a path for the plasma bolts. The Captain ordered the crew to aim at the weapons and they succeeded in securing some hits, but the replicator was apparently following the same strategy, aiming its plasma bolts at their laser batteries. After a full turn, the replicator resumed the same attitude in its course towards the moon, with all shields up.

"Damage report", the Hunter ordered.

"All our batteries are out", Beaumont answered. "It aimed at our weapons. There is no other damage, but all our weapons are gone".

"It clearly realized that on the moon there are many more organic lifeforms than on this ship, and it decided to deal with them first. It neutralized us so that it could act without being disturbed, but it will undoubtedly come back to us later. And unfortunately in our current state we can do nothing to protect the colony", the Hunter concluded. But now the replicator was starting its orbit insertion maneuver.

Since the batteries had stopped working, Mike had been studying the readings he was getting from the instruments on his console. He had an idea, but wanted to check whether it was workable. Just as the Hunter stopped talking, Mike had reached the conclusion that he was right. "Sir, it is not yet time to yield. From what I get from my instruments, the forward batteries are still working: those plasma bolts blew the control circuit's transmission lines. We are unable to control them from here, but the batteries are still working".

Beaumont was skeptical and, above all, was not happy that an outsider, particularly one who had been on board for such a short time, should interfere with what he considered his field. "What's the difference?", he said with an irritated voice. "If we cannot operate them, it is just as if they don't exist any more. The only thing we can do is to make a quick getaway from here, and have the required repairs done at the nearest space station".

"Sir, I can get outside in a space suit and reach the front tower. Once in the tower I can operate the battery by manual control", Mike went on, talking directly to the Captain. "I suggest that we should enter the same orbit. I am sure I can strike a hit through an opening in its shields as soon as it starts bombing the surface".

"No, I appreciate your bravery—I should say your recklessness—but I cannot allow you to do that. You will have no protection at all once you are outside: I cannot allow anybody outside during a battle. We must give up: that colony has no chance".

"Sir, you said that the replicator is now only concerned in attacking the colony, and is not paying any attention to us. I can sneak into the front tower without it realizing I am there and then strike without warning. We will take it by surprise, and that will give us a definitive advantage. We cannot give in now", Mike, said, getting up and starting to seal his space suit.

"You may well be right. We can try your strategy, but you need to be fully aware of the danger" the Hunter answered. Then he added: "If you succeed, I'll double your share of the reward".

"Thank you, Sir. Now I had better go", Mike concluded, walking toward the airlock. As he left the bridge, he saw Beaumont getting up and leaving his station, which was in any case useless now because the weapons could no longer be controlled from the bridge. He also heard the Hunter giving an order to turn the ship so that the path he had to follow outside was under cover. When he got to the airlock, he sealed his suit, put on a personal maneuvering unit and actuated the controls that opened the outer door. As soon as the door opened, he didn't wait to be outside before switching on the cold gas jets of his maneuvering unit, and he shot out from the hull like a bullet, aiming for the front tower.

Passing dangerously close to the thousand antennae and objects that stuck out from the hull of that type of ship, he reached the tower in a couple of minutes. He entered the tower and sat down at the console where the manual controls were located. He knew perfectly well how they worked: he had fixed so many towers of that sort. He had often wondered why they went on putting manual controls in weapons towers, when nowadays they were always remotely operated from the bridge. He remembered that on many occasions, while taking a break from working on that sort of tower, he had imagined himself there, sitting at the control console, in a space suit, shooting at a replicator. Well, now he was there in reality, and he had best concentrate on his job and take careful aim. If he didn't hit with the first shot, the replicator would return fire and he was in the most exposed spot. While he was nervously testing the aiming controls, he realized that the ship was entering orbit and rotating so that he was directly on the side towards his target.

His breath was taken away. He knew the size of a replicator and, moreover, he had seen this one on the screen on the bridge just a few minutes ago, but having it there in front of him, close-up, and seeing it directly and not on a screen was really scary.

The Hunter's voice brought him back to reality. "Get ready, we are entering a slightly lower orbit, so that the opening in the shields will be exactly in front of you when it starts dropping the first bombs". And then he added: "as soon as you open fire, I will launch all our torpedoes through the openings in its shields. If they strike, cover your eyes".

"We will have a nice firework display, Sir", Mike answered, trying to hide his uneasiness. He kept his lasers aimed at the spot where the bombs would be ejected. Then everything happened in an amazingly short time: he saw something coming out of the lattice structure that formed the hull of the replicator, and fired. The beam of light beam got the bomb dead center just as it was leaving the hull, and the heat made the ignition charge blow up, starting the nuclear reaction. As the bomb exploded he carried on firing. The hull of the replicator moved sideways and part of its side exploded, sending fragments all around. Mike saw the torpedoes reaching the hull, but also that the replicator's weapons had started to return fire with huge plasma bolts. Without even realizing what he was doing, he raised all the tower's protective shields and went on shooting blind. After a few seconds he felt the hull shaking as it was struck by fragments of the replicator and perhaps also by the plasma bolts, and he tried to relax.

He heard the voice of the Hunter shouting, trying to be heard over the deafening noise on the bridge. "Well done, boy. Struck dead in the middle. It's gone". Then he added: "engines at full power, we must get out of this hell-hole before the debris hits us".

12

All the noise coming from the radio from the bridge was in sharp contrast to the unbelievable silence that surrounded him. The hull was shaking and swaying in all directions, but in the vacuum of that depressurized zone all was silent.

Mike shut his eyes and tried to relax. He realized that his mouth was dry and his hands were shaking. He remained at his station, and slowly his tension drained away. When he realized that the ship was no longer vibrating and that the shocks on the hull had ceased, he lowered the tower's outer protection shields and left. He slowly moved toward the airlock, realizing that some areas of the hull had been depressurized.

'We got some heavy blows', he thought, and, after entering the airlock, started to walk in the direction of the bridge.

When he entered, he saw that Lakshmi Kimura, who also acted as physician, apart from being the life-support specialist, was bandaging the arm of one of the men. Three other people were sitting on the floor waiting their turn while O'Connor was lying on a stretcher: his eyes were open and he was speaking slowly, but he looked in quite a bad shape.

He went up to the Hunter, who was standing behind Liza Chen, who was sitting at her communications console. The same woman who was taking care of the colony's communications from the colony appeared on the screen.

"Captain, we are deeply indebted to you. If it were not for your intervention … Did you suffer any damage?" she asked.

"Yes, we took some hits, and five of our men were wounded, one in a fairly serious way. Do you have a hospital on Ceres?", the Hunter asked.

"Yes, of course. Bring them down and we will treat them", the communications officer answered.

Just at that moment an older man appeared on the screen. He introduced himself: "I am the mayor of this community, my name is Cheng Jian". Mike immediately realized that his appearance was typically Chinese: with that goatee he remainded him the copies of ancient Chinese paintings that he had seen in the headquarters of STSE Corp, on the main planet in the system. Certainly such marked racial characteristics were now rare, but in this case there it was not surprising: STSE recruited its staff and managers for their agricultural colonies in rural Chinese areas, which were not greatly involved in the homogenization of humankind that had followed its expansion into space.

"All of you come down. Tonight we will have a big party in your honor. It is the least we can do after what you did for us", the mayor added.

The Hunter looked round the bridge. "Thanks, my people will be pleased to spend some time on a planet: we have been in space for months. But I will remain on my ship".

It was clear that the Hunter didn't want to leave the ship that was now almost a wreck, to prevent anyone from getting on board and claiming salvage rights on an abandoned spacecraft.

"We have a prisoner: I would like to leave him at your colony. I must ask you to keep him in custody until a police ship arrives", he added.

The man asked whether he was a dangerous criminal, but when was informed he was Father Dubois, the mayor accepted without hesitation.

Mike waited for the conversation to finish, than he approached the Hunter and said "Sir, I can remain on board. I don't like much planets were one is out in the open, and then I could spend my time making some repairs. I can prepare a list of the most urgent spares and send it to you before you return. If you can get anything useful from the colony, we can start immediately. We can make some repairs on our way". The Hunter realized that Mike would have to act as chief engineer until O'Connor could return to duty and was glad of the offer. They could save at least a day and they had no time to waste if they wanted to get the other replicator before it could leave its asteroid and create havoc in the system.

"Thank you, I appreciate your offer. If you can send down that list by tomorrow morning, we will save a lot of time".

Mike started work immediately, while the others were getting ready to leave the ship and move the wounded and prisoners onto the shuttles. First of all he checked the outer skin for damage and re-pressurized all the compartments that had suffered damage. Then he started checking the engines.

He worked until late in the evening, and when he went to bed he was quite happy with what he had done. The list of required spares was almost ready, and he decided that they could leave that place very quickly provided it was possible to find all the parts in the colony.

The idea of spending the night alone on the ship in a system where there were replicators was quite frightening, but he was so tired that he fell asleep immediately.

The following morning he started work early, and by ten the list was ready. He broadcast it to the colony and went back to his repair jobs. In the early afternoon the Hunter sent him a message saying that they were ready to leave the moon, escorted by a small cargo ship with the spare parts and a load of fresh agricultural products.

The shuttles arrived first. As soon as they were moored in the hangar, the cargo ship docked outside and transfer operations started. The colony had a good inventory of spares for civilian ships, and they had found almost everything that was needed to fix the engines and the other systems except the weapons.

After checking what had been brought on board, Mike went to report to the Hunter: "With what you found we can fix the engine room and life-support systems, but there is nothing we can do about the weapons: we must get to a space station. The closest one stocking what we need is New Shanghai".

"Yes, so you can see that girl …", then, noticing Mike's expression, he added: "I didn't mean that you wanted to take us there simply for your personal interest. I too believe that it's a good choice. When can we leave?"

"If we leave as soon as the basic work is done and we do the remaining things on the way, we can leave orbit within six hours, Sir".

"Well, I give you authority to order anyone you think you need to help you."

He was about to leave the bridge when he stopped and asked "How is Mr. O' Connor? I didn't see him when you got back".

"He is well, but he won't be back. He resigned this morning", the Hunter said curtly.

As he went toward the door, Romero stopped him: "It was a bad idea of yours not to come to the party last night: there were a lot of girls, eager to thank us for what we had done for them. Not like those RGs that your friend had …".

Mike didn't answer and left, taking the corridor leading to the engine room to start work on the engines. Actually the work needed more than seven hours, and finally they left orbit, setting course towards New Shanghai.

As soon as the ship was on course, Mike got a call from the Hunter who summoned him to the wardroom, saying that he had to tell him something important. He realized it was almost eleven: he had worked the whole day without a moment's rest and now he would have preferred to get some sleep, but he had to do as he was ordered and so he immediately went to the wardroom to wait for the Captain.

It was only when he sat down, with nothing to do except wait, that he realized how tired he was.

The Hunter arrived after about twenty minutes, with a folder full of papers in his hand.

"Sir, you see that you finally got your seventh replicator. And that you have earned your place in the order of Sagan", Mike started, as soon as the Hunter came in.

"Thanks to you, boy. You must get the credit for that, even if in retrospect we must say that we took a big risk. Anyway, everything went well and nobody will censure us for it".

"Sir, were you speaking seriously when you said you would double my share of the reward?" During the day Mike had several times thought about those words.

"Of course. I never speak lightly about such things. And you earned that money. But there is more".

Mike didn't know what to think. He could think of many questions, but remained silent, waiting for the other to explain to him what he meant.

"I told you that O'Connor resigned, didn't I?"

"Yes, sir. Is he in such a bad state?"

"No, don't worry, his wounds were not very severe. For some time I have been watching him, and I could see he wanted to quit. You know, time stands still for no man. And at a certain point you start thinking that the small amount of reward money is not worth risking your life".

"Sir, it's not a small amount. What I earned on this trip is more than I could save my whole life working as a technician …" Mike interrupted him.

The Hunter didn't show any irritation for his interruption. "Sure, the first time you feel like that, but then, after several rewards are in your pocket, you start wondering whether it is worthwhile risking your life once again. And O'Connor just decided that it isn't."

"But you are not thinking of quitting. You will carry on hunting those bloody machines!"

"Of course, but often I ask myself why. I told you that you cannot hate those machines. Then, why go on?"

"Because it is our duty to get rid of them. You saw what almost happened to that colony. There are thirty thousand people on that moon, and nobody would have survived". Mike looked as if he really believed what he was saying.

"It's normal for you to say such things at your age. Anyway, I am happy that you think so. Because I meant to ask you whether you feel ready to take Mr. O'Connor's job."

Mike didn't immediately grasp what the other meant. Or rather, he couldn't believe what he was being told. "You mean I could become chief engineer on this ship?", he asked, realizing he was almost stuttering.

"Sure. Only temporarily for now: I'm hiring you for this hunt, until we get that replicator. Then, if we both are satisfied with each other, that job is yours. Is that all right, boy?"

It was only then that Mike realized what those words meant. He just answered "Yes, Sir", trying to hide his emotion.

The Hunter opened the folder he had brought with him and took out a sheet. He gave it to Mike and gave him a pen. Mike hardly glanced at the paper: it was a temporary contract, with all the financial details. Actually he didn't need to read it: he wouldn't miss that opportunity even if he had to pay for it. And then, if they could get that replicator, the chief engineer's share of the reward was more than years of his previous wages.

He signed the paper. The Hunter took it back and said "Good hunting, Mr. Edwards".

"Good hunting, Sir" he answered as he left. It was only after he was outside the wardroom that he understood that the other had called him Mr. Edwards instead of 'boy'. Now he was an officer on a starship! And on the Hunter's ship!

13

Mike concentrated on work during the six days they took to get to New Shanghai. He repaired everything that could be fixed in deep space with the spare parts they had obtained on Ceres, and by the end of it, the ship was in fairly good condition, except for the weapons. Actually, everything was working, except what was needed for hunting.

In the evening of the sixth day they docked at New Shanghai. Nobody from the maintenance department was around so late in the evening, but Mike had planned all the work for the following day by radio, and so he was expecting to have a full day off, to spend on the station, if possible with Ann.

As soon as the passageway was connected he left the ship and, leaving the docking zone, he saw Ann waiting for him: clearly Joe had told her the arrival time. She ran to meet him and hugged him. They remained there a few minutes, unable to speak. "At long last you are here. I was so scared when I knew that had you met that replicator, particularly because of the way you destroyed it", she said.

"We couldn't let it go. There are thirty thousand people on that moon: if we hadn't got it ..." Mike stopped: he felt he needed to change the subject. There would be time to speak about replicators and about the hunt, but now it was time to speak of other things. "Ann, I have to go to Accomodation: I no longer have a room here on the station. Will you come with me?", he asked.

"You don't need to go there now. Tonight you can stay in my room. And, in any case, now it is too late and nobody will be there".

He had not dared to hope that. "Of course, I'd love to". He put an arm round her shoulders and they walked silently towards her room.

When they were in her room, he sat down on the only available chair and she sat on the bed. "Ann, do you realize I am a rich man? That we are rich, I mean. You and Madame can leave this hole and go back to a normal life!" He got up and hugged her again.

"Were you really serious when you spoke about marrying me?" she asked.

"But of course, as soon as possible, say tomorrow", he answered. Then added with a laugh: "You don't think I'd leave you here, doing what you were doing before, especially now, with Romero around. On the ship he told me I had deceived him by suggesting to him that he should take that RG. He said that next time he would have the human I had prevented him from having".

"And what did you answer?"

"That he could go to hell before that happened", he answered, slightly twisting the truth. Then he added "What about giving me another demonstration that Romero was right when he said I cheated him?"

She laughed. "And why do you think I told you to come here tonight? After all, if it was just for sleeping, you could have remained on your ship …"

Mike got up and, looking at his watch, said "But before that, I would like to speak to Madame to apologize. I am sorry I couldn't speak to her after you told me your story."

"Of course. She wanted to come with me to the docking area, but then she decided that we would prefer to be alone. I think she is getting her RGs ready for the crew of your ship. Last time it went quite well, also thanks to your advertising".

Mike got up. "I will be back within ten minutes", he said going to the door. Then he turned back and asked: "do you still have one of those RG dresses?"

"But of course … but from now on I will wear them only for you", she said with a smile. "But now go, and don't keep me waiting."

"Ten minutes, no more", he answered as he went out. Madame's room was just a few meters away.

As soon as he knocked, the door opened. "Good evening, Ms. Donovan", Mike said, going inside.

Madame was standing, doing the usual maintenance operations on Lulu, who was standing close to her. The other RGs were standing there as well: clearly she had just finished checking them too. "Good evening, Mike", Madame answered.

The four RGs noted his presence and gave him very professional smiles, as they had been programmed. "Good evening, Sir", Lulu said in an inviting voice. Mike looked at her and noted she was really beautiful. He thought with pride that her voice was beautiful too, very sexy, and congratulated himself on his good work.

"Girls, wait in the other room", Madame ordered them.

Mike looked at Lulu as she left and thought that next time he adjusted her program he would increase the oscillation of the hip actuator by a couple of degrees. Perhaps the effect would be rather indecent 'but after all, RGs are not meant to simulate princesses', he decided. Only after ordering the robots to leave did Madame realize what Mike had called her.

"It's so many years since anyone has called me by my proper name", she said. "Here I am just Madame to everybody".

"I didn't even know your name until, just before leaving, Ann told me your story ", Mike said. "I came to tell you I am sorry about everything people say about you here on the station. But why didn't you tell people how things are?"

"And what should we have done? Put big boards on our doors saying 'we are not prostitutes'? And there again, all appearances were against us. We actually got a living from the work of these RGs, and from that misunderstanding. Even tonight, as soon as I knew that the Hunter's ship had docked, I started putting them in order. But soon we will go away … or at least I will go away, since now that Ann has found you, she will make up her own mind. Or rather, she will do what you decide together."

"Yes, Ms. Donovan. I have asked her to marry me and I hope she will make a decision tomorrow. I think Ann will ask you to leave this station with her … now there will be no problem. Don't worry about money: after all you did for Ann, we owe you that. You will be able to go to a safer place, sell those robots and be Ms. Donovan again".

"Yes, we have had so many adventures together. I know many think Ann is my daughter. She isn't, but in all these years she has practically become one. But sooner or later daughters must leave and lead their own life."

"Don't misunderstand me. Ann has quite an affection for you, and will always feel the same. And I, too, owe you a lot for what you have done for her."

"Thanks, Mike. But all I want is for her to be happy. And from what she has told me, I am sure you will succeed in making her so. Now go, she is waiting for you. If you remain here any longer, she will think I have rented out Lulu to you. She is so jealous of that robot" she ended with a laugh.

He said goodbye and left. Shortly afterwards he entered Ann's room. She was waiting for him, kneeling on the bed.

"You are beautiful, Ann", he said, as soon as he saw her.

She slowly took off her dress. "Now come here. I have waited for you for such a long time."

14

The following morning Mike woke up quite early. He sat up in bed and pulled the sheet aside. Ann was still asleep and he started to stroke her gently. Then he bent over and started kissing her lightly all over her body. He carried on for a few minutes and then stopped, still sitting up in bed. She slowly opened her eyes and with a smile said "Do you really have to stop?".

"How long have you been awake?", he asked.

"Since you started", she put her arms about his neck and pulled him down to her. "Do you realize that this is the fourth time tonight?", she said, with a laugh.

"Going by the clock, we can say it's the first time for a new day", he countered.

At ten o'clock they left her room, walking hand in hand.

"I'm really hungry. Let's go to Steve's and have some breakfast", Mike said.

"I'm hungry too, but we could have stayed in bed a bit longer", she answered.

"I had an arrangement with the maintenance boys: they already know what to do. I don't need to be on the ship before mid-afternoon. Let's have breakfast, then we can go back to your room", he replied.

Few minutes later they were sitting at a table in the cafeteria, which happened to be empty at the time.

Steve came out from the kitchen as soon as he saw them. He greeted Ann with a nod and then, addressing Mike, said "Good morning, Mike. I was eager to see you. Last night we wanted to come to the docking area to wait for you, but when we saw Ann we thought that it was better to leave you alone and to wait until this morning."

"Who do you mean by 'We'?", Mike asked.

"Why, Joe and myself, of course, but there were lots of others. You can't imagine how many friends you have on this station now that you are so famous … I set up a pair of big screens here in the corridor and when the news carried the story of your hunt for that replicator in orbit about Ceres, this place was full of people. We were all here rooting for you … just like the crowd in a sports stadium", Steve replied.

"Steve, we are hungry. Can you send us your waiter, so that we can order breakfast?", Mike asked, changing the subject. This talk embarrassed him, and he realized Ann was getting nervous. To calm her he put out his arm and placed his hand on hers.

"I don't doubt you are hungry: after all the exercise you must have had last night …", Steve replied.

Seeing Ann was blushing, Mike stopped him with a stare.

"Look", Steve continued, "Anyone only has to see you to understand how things stand. And Ann was so worried about you … Why did you leave that priest on Ceres? He may be a heretic, but he should be able to celebrate a marriage, shouldn't he?"

"For that we don't need the heretic: the captain of the station can do it, although he must be out of practice: when was the last marriage celebrated on this station?" Mike asked.

"Well, you get married and I will lay on a nice dinner. I will invite all those who wanted to welcome you at the docking area yesterday, which, in practice, is about half of the residents".

"I would like to see how you could do that, with that waiter robot that keeps going to pieces …", Mike joked.

"Generally it works well, when you don't sabotage it. Then we can ask Madame …"

"Ms. Donovan", Mike corrected.

"Well, Ms. Donovan, if that is her name, to let us borrow her RGs to wait at table. Don't say that you are unable to write a few lines of code to adapt their programming. And with their looks they will make everyone happier …"

"Look that we will take all that as read. We were just thinking of getting married", Mike interrupted him.

"And I'm not joking. I will provide the dinner. So you can say goodbye to this rathole in a great way. With all the money you have now, it won't be long before you move away from here. I imagine that you will go somewhere on the other side of the colonized zone, to one of those systems that will be safe from replicators for a few centuries. And you will start your own maintenance business. I can see it now: 'Edwards Enterprises—starship maintenance!'"

"That's a good idea, Steve", Ann interrupted him. "Starships and robots". Then, looking at Mike, she added with a laugh "Obviously, all robots except RGs. You will not be touching them any more".

"Yes," Steve went on, "but with the money of the rewards you can buy your own space station, as company headquarters. Perhaps on a main route, not in the middle of nowhere like this hole. Come to think of it, you know what I will do? If you grant me the license for a restaurant, I will come too. With all the ships that will dock there I will get rich too. Not like here where you get one ship a week, if you are lucky".

Mike pointed at the robot which had been standing at the table for some time. "I think we had better stop this foolish talk and get on with ordering breakfast".

Steve went away laughing, and they ordered a full breakfast.

"Steve's idea is not so foolish", Ann said. "The idea of moving to a safer place and using all that money you've got to start a maintenance company is a good one."

"The idea of buying a space station obviously is nonsense: the reward money is a lot, but not that big. And then …", Mike was interrupted by the robot, which began to serve breakfast.

They started eating. "And then, what?" Ann asked, since he said no more.

"And then I cannot leave this system now". He paused, then went on: "Tomorrow, if we can finish maintenance operations, we have to leave to look for the replicator that is hiding on that asteroid."

"But the Hunter knows where it is. You have no reason to go now", Ann said.

"I am the chief engineer on that ship. Just acting chief engineer at present, but I signed a contract for this hunt. And then I am sure that, if I don't do something stupid, I will be confirmed in that role. Think of it: I am chief engineer on the *Morning Star*, the Hunter's ship. If somebody told me just a month ago that I'd be that, I would never have believed him".

"You signed what? How could you do that, after deciding to marry me? How could you decide something affecting our lives without even telling me?" Ann was now speaking in a low voice, as if she wanted to prevent other people from hearing what she was saying. She was getting really upset.

Mike had known that she would not like it, at least initially, but he had not expected a reaction like that. "But, Ann, I must do it. Those replicators are destroying everything. They destroyed my family, they spoiled your life. We must stop them."

"There's no sense in doing it like that. Governments must accept their responsibility, they must get organized. It's pure madness to leave everything to private individuals, to bounty hunters that each operate individually. To fight against replicators we need organized fleets: a single ship has little chance of surviving a battle against a replicator", Ann went on.

"You are right, from a strictly technical viewpoint", Mike answered, "but you know the situation: the only planet with a true government is Earth, and Earth will never deal with this problem until replicators are a direct danger to its survival. The colonies are small and poorly organized; they will never put together a combined fleet. It is already somewhat of an achievement that they agreed to fund the rewards. No, hunters are the only possible solution, and you know it as well as I do."

"And then what Steve said makes even more sense. Our only chance is to move to a safe system, and leave hunters to their folly".

"No, Ann, I must fulfill the contract I signed with the Hunter, at least for this hunt".

"And then go, you and your Captain Ahab …"

Mike interrupted her "The Hunter's name is Hillman, not Ahab".

Ann burst out laughing, even though she was upset. "You never read Moby Dick, did you?"

"No …, what's it about?", Mike asked, happy to change subject, even if just for a moment.

"It deals with a mad man, Captain Ahab, who is the captain of a whaling ship in the nineteenth century, on Earth. He is so obsessed with hunting a particular, huge whale that in the end he dies, taking all his crew with him. All but one, who escapes using a coffin as a raft".

"A cheerful book, by the sound of it. No, I've not read it, and I have no intention of reading it. But the Hunter is not mad, and he has already destroyed seven replicators. I am sure he will survive a good number of future hunts", Mike concluded, taking the subject for granted.

"Do as you like, but I have no intention of staying here, waiting for a hunter who has every chance of not coming back alive every time he goes away". Ann, too, thought she had settled the argument once for all.

"No, Ann", Mike tried again, in a subdued voice. "Officers can take their wives on board. You can come with me. You will see that living on a starship is no worse than on a space station like this. And then, you like astrodynamics; in time you could become a full member of the crew. By the way, you will then be entitled to your share of the rewards", he concluded, trying to smile.

"Yes, you can take any object with you, even a wife. The Hunter tried that as well, and we all know how that ended …"

"No, the Hunter's wife was a member of the crew: she was the ship's physician and communications officer", Mike tried to say, in an even lower voice.

"No, Mike, I have already had my encounter with a replicator. And I swore to stay away from them. You must make a choice", she ended, sounding like a person who was about to cry "Either you give up the hunt or you leave me to get on with my own life".

Ann got up and started to move towards her room. Mike grabbed her hand. "Please sit down. Let's talk", he said in a voice that he tried to keep calm, but which betrayed all his emotion. "At least finish your breakfast. Then we can talk more easily".

Ann pulled her hand free "You eat, if you feel like it. And there is nothing to add. You must simply make a choice". She spoke without turning: she didn't want him to realize that she was crying.

15

Almost immediately Steve came out to the table. "What happened? I saw Ann going away."

"What happened is that you saved the money for the dinner. We quarreled: she doesn't want me to leave with the Hunter"

"Well, what do you expect? She's right: you got away with it once, you cannot expect to be so lucky every time. Being a hunter is not a job like any other, it means you have to risk your life every time. Why do you think they give you all that reward money?"

"But I can't. For years I have been dreaming of becoming a hunter, and now that I have succeeded she cannot ask me—none of you can ask me—to withdraw."

"I'm not asking you anything. She is a sensible girl, and has already had one encounter with a replicator. You wanted to get your revenge for what happened to your family. Now you've got it, and in doing so you saved thirty thousand people. You have become famous and awfully rich. Don't be stupid, go after her and tell her you have changed your mind. This is your big opportunity: I was not joking when I told you to open a maintenance business. You are a clever guy, and in a few years you will become a rich and respected businessman, with a beautiful wife, a number of children and a lot of money. If only I were in your place …"

Mike had to admit that Steve had a point, after all. "I cannot betray the Hunter: he needs me and I cannot just leave, leaving him without a chief engineer".

"Of course you can. Tell him you have changed your mind. You will pay a penalty for breaking your contract—but for you now that will be peanuts. He can find another chief engineer and everything will be settled."

"You don't understand, Steve. Not only I would betray him, I would betray myself, my dreams, the promises I made to myself when my family was destroyed."

"Bullshit. You have no duties towards anyone, except to yourself and Ann. And among those duties there is not getting killed by a replicator to get a little money. Well, it might be a lot of money, but it is money you don't need, now that you are so rich".

"Steve, you don't understand. You are only a coward who's giving cheap advice, and unsolicited advice at that. Mind your own business, and I shall mind mine. What is it with you all? Did you gang up together to get me doing what you decided? Don't you realize that I can decide by myself what I have to do?", he replied, getting up and walking off towards the docking area.

He went onto the ship and to his new cabin. Since he had been appointed chief engineer he had moved from the small room in the lower bridge, were he was lodged when he joined the ship, to a large cabin on the officer's deck. He threw himself on his bed, He really felt like crying.

He realized that his dreams were contradictory, now that they were about to come true. He couldn't become a hunter and marry Ann at the same time. And it was hard to make a choice. He tried to convince himself that this was the result of too much luck; but now all he really wanted was to get back to being a simple maintenance technician on a forgotten space station, with many dreams that would never come true and no difficult decisions to take.

Slowly he relaxed. He couldn't miss the chance of becoming a hunter. He felt that in time he could become the commander of a ship, of his own ship. To become famous, and a renowned hunter. When the Hunter retired, he perhaps would inherit the name. And those cowards, who had never been in a battle, were exaggerating the risks. What do Ann and Steve know about hunting? Only what they have got from the news, from what had been said by journalists and writers, who are only interested in sensationalism.

'The replicator that will overcome Mike Edwards has still to be built', he thought.

But perhaps it was worthwhile trying again to convince Ann. He switched on his screen and ordered the communication system to call her.

Ann answered immediately. She was in her room, and her eyes were red. "So, Mike, what have you decided?", she asked immediately, without even saying "Hello".

"Ann, we must talk. It must be possible for us to settle things without having to take such drastic decisions". He stopped for a moment and then went on "I don't want to lose you, Ann".

"Neither do I want to lose you. But I don't want to be the wife of a hunter. What did you decide?"

"I cannot leave the Hunter in this way …"

"And then follow your Ahab, Ismael. And don't forget to take a coffin with you", she said, switching off the communication channel.

Mike remained there, staring at the blank screen. He didn't understand the last sentence, which felt like an obscure threat. 'That bloody book', he thought. 'It's a good thing I am not superstitious'. He stretched out on his bed, trying to relax.

It was now clear that his attempt had failed: Ann would not let him convince her so easily. But he could not give up without trying again. He waited some time to pull himself together, then left his cabin to see how the repairs were proceeding. He met none of the crew of the ship; many were in the station where they had rented rooms, others, like the Hunter, had remained in their cabins.

He needed to talk, and the only person on board with whom he was friendly enough to speak about his problems was the scientific officer. He tried knocking at his door and, to his surprise, the door opened.

"Come in, Mike" said Campos. pointing to the only chair, without even raising his eyes from the screen in front of him. "Thank you, Takashi", Mike answered. "But if you are busy I'll come another time", he added.

Campos turned toward him. "I thought I was alone on board, apart from the Captain, naturally. Why aren't you on the station, with your friends?"

Trying to overcome his emotions, Mike told him what had happened.

"I understand, Mike", Campos answered, when he had finished, "but you must get used to that. A hunter is a lonely man. Everyone on board is lonely, and those who tried to have a normal life, like the Captain, had a whole lot of problems."

Seeing that Mike was about to speak, he gestured for him to keep silent, and went on: "The distances between star systems are huge, and a hunter wanders from system to system. Living on a ship like this is not like living on a cargo ship; you cannot compare this with what you experienced when you were a boy. A cargo ship may be a family enterprise, a hunting ship cannot. Why do you think the rewards are so high? All hunters think they can follow this life for a few years, and put together enough money to be able to retire and lead a normal life", he said.

"But then, is it worthwhile to go through all those hardships just to earn that amount of money?" Mike asked, realizing that he was really asking that question of himself.

"I don't know. Each of us has his own motivations and his illusions …"

"So then, why are you on this ship?", Mike asked him directly.

"Well, mine is a peculiar situation. I worked for a famous university, on Earth …", he stopped, looking at the other's amazed expression. "Yes, on Earth. But you must not think that Earth is such an idyllic place, as people from the frontier tend to do. Earth is not the paradise you may imagine. Anyway, I had a nice life, I cannot deny it. But I was fascinated by replicators: they are the only evidence that we are not alone in the Universe. Any life that we have found is just bacteria and quite simple lifeforms. But somebody must have built the first replicator …"

"And so you are here neither for the reward money nor to get your revenge on those machines, but to study them …", Mike interrupted him.

"In a way, yes. At one time the Hunter came to Earth and suggested that my university should study replicators. He proposed funding a research project. By then he was already quite rich and he could afford it. I believe that his ultimate goal was to get help in understanding them to make it easier to fight them. I asked him if I could come on board as a scientific officer. At first he didn't take the idea seriously—after all no hunting ship ever had a scientific officer, like an exploration vessel. Then he realized that it made sense, and he not only allowed me to carry all the equipment that i considered to be useful on board but also accepted me with a rank of an officer, including my fair share of rewards. I left everything and I am still here …"

Now Mike understood why the ship had a scientific officer. "And since then, what have you found out?", he asked.

"A lot of things, and yet nothing at the same time. To be precise, nothing relating to what I was interested in: Who built them and why, where they come from, why they destroy any organic life they can find. I know as much as you about that, even if I have a lot of theories. But they are just hypotheses, without no evidence behind them. Every time we begin a hunt, I decide it will be the last one, that if I don't discover anything, I will go back to Earth to enjoy the wealth I have accumulated all these years. And then, at the end of that particular hunt I feel that next time will be the final one, that I cannot withdraw without finding what I am looking for. And then after so many years, there is nobody on Earth for whom my return would be of any worth.

When Campos had finished they remained silent for a while.

"May I spend some time in your lab? I know something about robots and I too am interested in understanding how those bloody machines work. After all they are nothing less than robots …" Mike asked.

"Sure, but you must not take it for granted that they are robots as we understand them. Even identifying them as Von Neuman machines, like most people do, might be wrong. Might they not be lifeforms based on metal and other inorganic substances instead of carbon and organic chemistry?"

"But wouldn't that mean that nobody built them … That they evolved like we did?", Mike asked, in disbelief.

"Why not? After all, for all we know, it might be like that …"

"Ja, but then father Dubois might be right … If they are not robots, they might be intelligent …", Mike said.

"What has their being robots, built by someone, to do with the fact that they may be intelligent, and above all conscious? The latter depends on what is needed for anything to be endowed with intelligence and self-awareness, and nobody knows what that involves. After all, we are always facing what Kant called the problem of the soul, which he said cannot be solved by using pure reason …"

It was the second time in a few days that someone had mentioned the German philosopher to him, a philosopher about whom he knew little more than his name. "Listen, I enrolled on this ship to hunt replicators, not to discuss philosophy", Mike interrupted him. He had taken a glance at his watch, and realized that it was getting late, and he had to check how the operations to get the ship ready for their next hunt were proceeding.

He got up, said goodbye to Campos and, as the latter turned back to his computer, left the room.

Mike looked for the head of the maintenance team, a second-class technician just as he had been until a few days before, and asked him to explain what they had done up to then. The batteries were almost battle-ready and the other systems were also almost operational. They would be able to leave at noon the next day.

At that point he was eager to go. If he was to remain alone, it was better to start his new life as soon as possible. At dinner time he decided to leave and go to Steve's.

16

As soon as he reached the canteen he sat down at a table.

"Well, Mike, have you made a decision?", Steve asked as soon as he saw him.

"Of course, the only decision I could make. I cannot quit hunting, now that at last I've got on a ship. For years I have been looking for an occasion like this, and you can bet that I won't throw it away for ..." he paused, looking for the right words, than concluded with a rush "for the whims of a scared little girl".

"Think better of it, Mike ...", Steve started.

"Enough with this bullshit" Mike said without letting him to finish the sentence. "I came here to have dinner, not to listen to your sermons. Send me that robot of yours, I am hungry. You nearly made me miss my breakfast and I have had nothing to eat since this morning".

Steve went away and sent him the waiter.

While he was waiting for the food, Mike tried to be cool and think. Perhaps he could make another attempt. He ate quickly, got up, and started walking toward Madame's room.

She let him in and met him with a smile.

"Did Ann speak to you?", Mike asked immediately.

"Yes, she came here this morning as soon as she left you. What did you decide?"

"What do you think I decided, Ms. Donovan? You all know full well that I cannot leave the Hunter like that. I have an obligation to fulfill and I have an engagement at least until that replicator is history", Mike answered.

"Then I believe that nothing will change Ann: she looks as if she really means it. To be frank, I am really sorry. You were such a nice couple. Ann needs someone who can offer her a future far from places like this."

"I was wondering whether you could speak to her, and explain the situation ..."

"So that's why you are here. I fear I cannot do anything for you. Ann has already encountered a replicator and her life was upset by that. There is no way out, Mike. You must understand that. Take her away, far from replicators, go to a system they will never reach, at least in your lifetime. You cannot ask her to be the wife of a hunter".

Mike lowered his head "If that's how things are, I don't think there is anything to do", he said getting up. "I want you to be happy, tell Ann that. But, please, leave here as soon as possible: this station is no longer a safe place, with at least one of those monsters out there ...".

"Thanks, Mike. We will go as soon as possible. And you too, try to have a happy life. I hope you will not regret this decision". Then, after a few seconds, she added "Good hunting: I think that's what you say".

"Thank you, Ms. Donovan", Mike said finally as he left.

He started walking slowly towards the ship. Now he realized that he was really alone, and that his life was at a turning point. Now his decision had been taken for good.

After a while he suddenly had an idea. He turned back and walked quickly towards Madame's room. He knocked at the door and entered as soon as the door opened.

"Have you changed your mind?" she asked, after greeting him briefly.

"No, Ms. Donovan", Mike asked. "I just wanted to ask you a favor."

"If you think I can convince Ann, I fear it is useless".

"No, it has nothing to do with Ann. I too think that nothing will change her mind", he answered. After a few seconds he added: "Would you please sell me that RG 46?".

"That's funny. Do you want to open a brothel on your ship? We need that RG. And I don't have the least idea of how much she may be worth".

"No problem, I will pay twice the cost of a new robot of the same model. I know that, after all the changes I have made, it is worth much more than the standard robot. If you like I can pay even more for it".

Madame was puzzled. "All right, you have a lot of money, and you nouveaux riches don't mind what you spend". She remained a few seconds without speaking, then she thought she understood. "No, Mike, thanks a lot, but we cannot accept".

"But why? After all you still have the other three, and with what I give you for the 46 you can easily buy a ticket to a safe place".

"That's the point", Madame answered with a smile. "It is very generous of you to give us all that money and to buy that robot to make us feel at ease, but we cannot accept. We managed in situations worse than this, and we can continue by ourselves."

"No, Ms. Donovan, don't misunderstand me. I don't want to give you alms. I really want that robot: I have worked so much on her and I would like to go on. We can say it is a research project in applied robotics".

"If you want to work on a robot, why not use a simpler one. It would cost you one tenth of the amount or even less", she answered.

Then again she thought she understood. "Perhaps that's not it. You are trying to make Ann jealous, so that she changes her mind, aren't you?"

"No, I am not that clever. I've no hidden motives. I'll buy that RG 46 because I am interested in an RG 46. And then I have given up with Ann. If the problem is money, we can settle for three times the market price".

"No, Mike, I am not trying to raise the price. Well, if you want Lulu, take her. However, she is not here now. That first mate of yours came this afternoon and rented her until tomorrow morning. He was quite upset and wanted Ann at any cost. But when he realized that Ann is not working any more—I told him that—he settled on Lulu".

"That is your business. Ann can do whatever she prefers to do. Tomorrow morning I will buy a crate, then I will come here, to dismantle Lulu and take her with me. In the meantime I will put in an order for the money transfer. Is that all right?"

"Don't worry, I trust you. But can you tell me why you want Lulu? And don't give me that spiel about applied robotics research."

"Because this way I can have a girl who won't try to have me doing what she wants", Mike answered.

Madame couldn't refrain from laughing. "But Mike, she isn't a girl, She is just a machine. Lulu does not exist: the only real thing is RG 46/G".

"Well perhaps you are right. But the work I want to do on her is not just dealing with the voice synthetizer, the actuators or the heaters. Father Dubois says that complex robots have a soul. Perhaps this is impossible—I don't know, I am no theologian—but RG 46 will really become Lulu. What does it matter if her bones are metal and her flesh is silicon? Her mind will become human".

"Be careful, Mike. Three hundred years ago Tipler said the same things about Von Neuman machines, and the aliens who built the replicators perhaps shared that view." But then she thought to herself that if he ever succeeded in his task and Lulu truly became a person, he would have the same problems with her as he now had with Ann. 'There's no point in telling him this now', she thought. 'If he ever succeeds, he will realize for himself.'

"Well, Ms. Donovan, Tomorrow morning at ten I will be here", Mike finished as he left.

17

The following morning Mike got up at seven. He checked how the maintenance work was proceeding and saw that everything was fine. Then he called the Hunter to report on the situation.

They met in the wardroom. Mike gave him a detailed account, listing all the work that had been done the previous day and those issues that were still to be finished. "The ship will be ready to move at two p.m.", he ended.

"Very well, Mr. Edwards", the Hunter replied. "Please inform all the crew to be on board by one. We will leave as soon as possible".

"Sir, I ask for permission to bring a crate of personal stuff on board", Mike asked, quite ill at ease.

"No problem, all officers are allowed to bring personal belongings on board. What are they?"

Mike knew that he had the right to such belongings, but that the Captain had to be informed of their nature and that he could forbid objects that could be dangerous, injurious, or liable to be detrimental to the normal operation of the ship. Each captain interpreted this rule as he thought fit: some prohibited spirits, others objects linked to gambling, some prohibited almost nothing and others almost everything. Blushing he had to admit: "It's an RG, Sir".

The Hunter stared at him with an expression halfway between amazement and disgust. "What? Do you intend to turn my ship into a brothel? You'll not do that".

"But, Sir, I don't mean to rent her to anyone. I'll keep her in my cabin … It's just for personal use".

"For personal use? Not even Mr. Romero ever did a thing like that. He at least waits until we reach a station".

"Don't misunderstand me, Sir. I want to work on that RG to assess the possibilities of artificial intelligence. I have been modifying that robot for months and I just mean to go on with that work in my free time."

"And you want to sell me the idea that you are using an RG to study artificial intelligence? And since when is anyone concerned with the intelligence of RGs?"

"No, Sir, you are wrong. RGs are the most complex robots, those that have a most sophisticated control system". He had come up with an idea that rationalized his desire to take Lulu on board, and he wanted to elaborate on that. "Since they are robots whose only task is to interface with humans, they are more sophisticated than others. So I decided to start with them in my research. You could say that the basic aim is to develop a new form of HMI". He saw from the Captain's expression that he didn't understand the acronym, so he explained: "human-machine interface."

"I know the interface that people use with those machines …", the Hunter interrupted him.

"Please, Sir, don't let appearances mislead you. The study of artificial intelligence will lead to a better understanding of replicators as well, something that is vital for hunting …"

"Well, bring that bloody RG on board, and stop bothering me with these idiotic justifications. But on one condition. She is never to leave your cabin and you will never speak about her to anybody. If I find that someone knows that an RG is on board, I will throw her overboard".

"Thank you, Sir", Mike said as he left.

It did not take him long to get a suitable crate and the tools he needed and then he went to Madame's room. Lulu was there waiting, on a chair. The robot had already been switched off, and in a few minutes Mike dismantled her and put her in the crate. He also took some of the usual accessories that come with RGs. The he asked: "Ms. Donovan, as I am here, do you want me to check the others one last time? Now that you no longer have Lulu, they will have to be fully operational".

"Thank you Mike, but now that we have all that money, we can get away from here on the first ship and I hope that we will never need those RGs again".

Mike told her that those RGs had allowed them to survive for years, and that it was wise to keep them working.

"You are right, nobody knows what life has in store for us", Madame agreed.

He worked for more than half an hour doing the usual maintenance operations on the three robots, then he said goodbye and left with his precious crate.

Once he got to the ship, he immediately stored the crate in his cabin, and went to the engine room to get ready for leaving the station.

At one he was on the bridge with all the others. They left the station half an hour later, setting a course towards the asteroid where they hoped the replicator was still to be found.

Later that afternoon, when his turn on the bridge was over, Mike went back to his cabin. First of all he reassembled Lulu and downloaded all the programs he had used to modify the programming of the robot onto his allocated space on the ship's computer, a thing that kept him busy for more than three hours. Then he started to check all the higher level programs, those that determined the behavior of the machine. It was there that he had to make changes, if he wanted to obtain responses that were more intelligent and less automatic. He stopped briefly only for dinner, and then carried on, seemingly without obtaining any interesting results.

At about eleven he decided that he was fed up with working. He sat on his bed and switched on all the robot's functions. "Well, Lulu, show me what you are able to do".

"With pleasure, darling", she answered. "Would you like some music?"

"Yes, please, but keep the volume low", he answered. He didn't know how good the acoustic insulation of the cabins was.

The room was filled with voluptuous music, while Lulu started dancing, undressing slowly. Mike was staring in fascination, thinking with pride that if she was able to move in that way it was because of his work. He realized that the designers of RGs were right when they said that, if well programmed, these robots could fulfill the fantasies of males better than true girls, since they where the projections of the very same desires. When she had undressed completely, she stopped for a short time to let him stare at her, then asked him "do you want me now?"

"Definitely, Lulu", he said stretching out on his bed. "Come on".

He had never really tested her, bearing Ann in mind. But now, that she was no longer part of his life, he did so, without feeling guilty.

When it was all over, he thought that Romero wasn't wrong when he said that, when you get to the real thing, she was not really that hot. He made a mental note of the modifications she needed, told her to switch off, and went to sleep.

The following ten days went by slowly. The ship was racing at top speed, aiming for the estimated position of the asteroid, which had been revised on the basis of the trajectory of the first replicator, the one they had destroyed while it was orbiting Ceres.

When he was not busy on the bridge, Mike devoted all his time to what he had described as his applied robotics project. At first, he tried to improve the robot's high-level cognitive functions, and initially obtained some interesting results. Soon, however, he realized he had come to a dead end. He could obtain reactions that mimicked human behavior, but if the situation became more complex he obtained responses that either had little meaning or, more often, were simply repetitive.

After a few days he went back to working primarily on the actuators. There the results were more satisfactory. After all that was just a control problem, although a complex one, because it involved a large number of sensors and actuators, but conceptually it was much simpler.

He tested her often, and found that the improvements were substantial. Several times he thought of having Romero try her, to show him what he had been able to do, but increasingly he found that he wanted to keep Lulu just for himself. He wondered whether he was getting jealous, but as usual he concluded you cannot be jealous of a robot, at least not until he had succeeded in transforming her into a real person. To keep the RG hidden in his room, ready to satisfy all his desires, made him feel like a sultan with a harem. Not much of a harem, with just one girl, but after all he could have had as many as he wished if he only knew where to store them.

From time to time he went along to the lab, to chat with Campos. He too was obsessed by a project. He was collecting all the available evidence about sightings of replicators and the rare broadcasts that had been received. The latter were just exchanges of data between two replicators, but he hoped to be able to detect their routes and their operational behavior by analyzing them statistically.

"It is more than thirty years since humans came in contact with those monsters, and we still know almost nothing about them", he told Mike, one day when he was particularly downcast.

"But isn't it possible we can understand something from studying their fragments?", Mike asked. "After all, when we blow them up something remains."

"Hunters do not bother looking for fragments. Initially, they hoped to salvage something that was worth being sold. They thought that there had to be a lot of exotic and precious materials in such machines. Soon, however, they realized that they could be neutralized only by destroying them completely and the explosions scattered their fragments over such a vast area that it was practically impossible to salvage anything of value. The time wasted in collecting the few remaining pieces was not worth their value, and now nobody bothers to collect them". Then, seeing Mike's disbelieving expression, he added: "The Hunter is different, he is not just interested in money, he also wants to understand our enemy. It was easy enough to convince him to look for fragments, but the few we got yielded almost no information."

He gestured Mike to open a big cupboard, and showed him some chunks of what looked like metal stored methodically on the shelves. Each had a label stating the location and the date of the battle and a few other details. "Some of them are just made of metal, alloys based on iron, aluminum, titanium or magnesium. They are alloys that are not very different from those we use to build spacecraft. Others are made of more sophisticated materials, some appear to be made of carbon nanotubes or other materials built using nanotechnologies. Overall, their technology does not look to be very much more advanced than ours, at least when it comes to material science", Campos said.

"In a way, their technology is less advanced than ours", Mike noted, "they do not have the warp drive and they take many years to go from one system to another …"

"Luckily so", the scientific officer commented. "They are already dangerous enough as they are, think what could happen if they could travel as fast as we do … By now we would already have been swept away", he concluded. Then he added: "Actually we cannot say we are more advanced than them. They are much better than us at artificial intelligence. We couldn't build machines able to replicate themselves, or to run a battle in an autonomous way. And in the field of weaponry they are more advanced than us. If we could only have those plasma jets …"

"It looks as if, technologically, we are almost even, so the outcome of each battle is uncertain", Mike concluded, thinking of what had happened in Ceres' orbit.

"Perhaps they are better than us in technology, after all. We have managed to survive because of our intelligence", Campos said.

"We should try to capture one of them in one piece—if that is ever possible", Mike said.

"Ja, if only it were", the other agreed.

Whenever he spoke to him, Mike had an urge to tell Campos about Lulu. He could perhaps have helped him with his project. But he never dared, because of the Hunter's prohibition.

Late in the afternoon of the tenth day the alarm rang, while he was in his room making his last attempt to improve Lulu's high-level control. He switched off his computer, logged out of the ship's mainframe, and ran to the bridge. An asteroid, still quite far away, was on one of the screens.

"Reduce speed and switch off all active sensors. Telescopes at maximum magnification", the Hunter ordered.

The image on the screen became larger, but it was still too poorly defined to see any details, in particular because of the bad lighting.

They closed in cautiously. "Analyze the infrared signature", the Hunter ordered again. The image on the screen changed; now the asteroid was all dark, except for a small spot, towards one side, that showed a dim image.

"What do you think, Mr. Campos? Could that source of heat be a replicator?", the Hunter asked.

"It's possible, but it is hard to say from here. However, I don't see what else would be hot enough on an asteroid of that kind", he answered.

"It might be a mining ship", Mike suggested.

"I checked with the mining bureau and no miner has claimed rights on asteroids in this area", Chen said.

"Sometimes miners don't claim rights on asteroids in remote areas. They don't trust the bureau, and fear that someone else would steal their rights", said Mike, who had spoken to many miners.

"Well, it could be many different things, such as a raider's ship or an outlaw's. But I believe that there is a good chance that we have found our replicator", the Hunter stated. Then he added: "Let's increase our distance, moving in such a way that we can later close in from the opposite side until we get within a few kilometers of the surface. We must take it by surprise", he ordered.

The maneuver took a long time, particularly because they remained on low power until they were completely hidden by the asteroid's bulk. Then they closed in quickly and in less than twelve hours they were less than a kilometer from the surface. Nobody left the bridge for the whole of that time, taking turns to take short naps at their stations.

The asteroid now filled the screen with its craters and its irregular shape. They started to fly across it slowly, following the irregular surface, to approach the zone where they thought the replicator was located.

They crossed over a large crater. It was different from the others, both because it looked recent, and also because it was full of debris. "Here we are", the Hunter said. "The replicator dug that crater to get the materials to build a copy of itself. Get ready to launch ten torpedoes and to fire the batteries. Aim at the zones that are in contact with the asteroid; I believe that there are discontinuities where the shields touch the surface". They went down even lower and continued to follow the irregular shape of the asteroid.

Suddenly, after flying over a mountain, they saw it in front of them: a huge replicator lying in the bottom of a crater.

"Fire at will", the Hunter ordered.

The torpedoes hit the shields in the area where they were in contact with the surface, causing them to saturate, and creating large holes in the shields. The laser beams then penetrated through the rents, directly hitting the replicator's structure. The huge machine was thrown up from the surface, revealing its underside, which was unprotected by the shields that had started to activate. The concentrated fire from the batteries first destroyed the shield generators and then triggered a series of explosions that shook the replicator. It split into two parts and then fragmented into countless tiny pieces.

18

Mike stretched out on his seat, closed his eyes and relaxed for a moment. This time everything had gone right. The replicator had been destroyed before it could strike back. The Hunter really was worth of his reputation. When he opened his eyes, he saw Romero getting up and leaving the bridge. Within a couple of minutes he was back with a bottle in his hand. "This time we must celebrate: this is the best victory for years!", he said in a loud voice. Mike looked at the bottle and realized that it was champagne, true French champagne. He had never seen a bottle like that: only a hunter would be able to afford it out in that remote planetary system.

"You are right, Mr. Romero", the Hunter answered, with one of his rare smiles. "But we will have to be quick; we must study what remains of that replicator. Perhaps this time we will get something that can allow us understand something more about our opponents."

Nobody dared to answer, but from their expressions it was clear that nobody liked the idea of starting work again before a good rest. Only Campos looked eager to start.

They drank the champagne quickly and then the Captain assigned tasks to each of them. "Mr. Edwards", he said to Mike, "You will study the images of the explosion in detail, and then go out onto the surface with Mr. Campos and try to locate the most interesting fragments. You have full authority to order anyone not engaged in more urgent work to help you to carry on board anything you think may be useful."

"Next meeting in the wardroom in eight hours. Everyone will report on what they have discovered", he ended.

"But Sir …", Romero started to say.

The Hunter gave him a withering look: "We will get some sleep when we have finished".

For more than half an hour Mike studied the images that had been recorded when the replicator blew up, trying to understand where the largest fragments might have ended up. Then he mustered a team of four men, put on his space suit and went out onto the surface.

Campos was already out, examining the crater that the explosion had produced, in the light of a number of searchlights that had been installed on top of the ship. As soon as they were outside, Campos called them. He had found a fragment some tens of centimeters long. They recorded its position accurately and took it on board.

They went on combing the crater and found several other pieces, At first sight nothing particularly striking, but they were certainly potentially interesting stuff. They took a lot of pictures and noted the position of each one. When they returned to the ship six hours later they had gathered several tens of kilograms of material, including several fragments more than one meter long. They took everything to one of the repair workshops and Mike and Campos started trying to identify where in the replicator each piece came from, starting with the largest ones.

The time went by quickly and they almost failed to realize that it was eight o'clock. They hurriedly collected their notes and rushed to the wardroom.

When they got there, everybody else was already there and the Hunter gave them a rather icy look. "Well, now that also Mr. Campos and Mr. Edwards are here we can start", the Hunter said, gesturing for Campos to begin his report.

"Please, before we start I must warn you about a rather worrying occurrence", Bhagwat stopped him.

"Go on, if you think that the matter is that important", the Hunter told her.

"I compared the mass of the asteroid, as computed in the miner's reports that Mr. Edwards gave us, with the mass that I measured", Bhagwat started. She was nervous and spoke quickly. "There is a lot of mass missing, much more than that needed to build the replicator that we destroyed close to Ceres."

"It is likely that this asteroid is made of materials poor in usable material, and that the replicator had to mine a lot of ore to get what was needed to build the new one", the Hunter interrupted her.

"I don't think so, Sir", Bhagwat went on. "I analyzed the minerals in the zone around the crater the replicator dug, and their composition doesn't justify such a large decrease in mass". Mike looked round and noticed the bored expressions of the crew. Clearly they didn't appreciate all these technical details. "Apparently the machine replicated twice", Bhagwat ventured to suggest, speaking very quietly.

"If it did, the second replica cannot be far from here. It didn't have enough time …" the Hunter commented, as if he was thinking to himself "And that might also explain why we destroyed it so easily. It was probably still on the asteroid because it was reconfiguring itself for spaceflight", he concluded.

"But then, if the first replicator we destroyed had broadcast the position of Ceres, the new one would certainly set course towards the colony. Perhaps we can get it before it reaches its target, if we hurry up", Kimura observed.

"You are right, we must not waste time", the Hunter said, leaving the wardroom and moving towards the bridge. "All hands to their stations, lift-off as soon as possible".

Within half an hour the ship was in space, on course towards Ceres.

19

When, after making all the checks required by their sudden departure from the asteroid, Mike was able to go to sleep, it was already four a.m.

The following day everyone settled into the usual spaceflight routine again, except for Mike and Campos who were very busy. Almost without realizing it, Mike neglected his 'applied robotics' research to work full time, together with the scientific officer, on studying the images of the replicator and the fragments they had collected.

It was not that he was not thinking about Lulu any more, but in practical terms he seemed to have given up trying to modify her control software to obtain a more human behavior. When he thought about the matter, he convinced himself that he was doing the right thing by saying that he had not given up, just that he had more urgent work to do. And then, after all, what Lulu needed was simply some adjustment to her actuators. She was actually beautiful and even in bed she was not that bad.

It did not take long for them to succeed in understanding where many of the fragments came from and for them to decide that some of them appeared to have come from the central control unit. When they had obtained some convincing evidence, they called a meeting with the captain in the wardroom.

When they described their results, they initially met a strong skepticism: after all it was the first time that anyone had stated that they understood anything about the replicators' architecture. It was clear, however, that these results were thanks to a unique stroke of luck, because they had succeeded in destroying one of them, not in open space, but sitting on an asteroid, something that had never happened before. The replicator was not moving and the fragments did not disperse in space, but were found in well-documented locations. Moreover, the video sequences of the explosion had been shot from a close distance and were quite detailed.

Owing to these unprecedented circumstances, they could identify some parts of the replicator that might be vital and which could be destroyed by a few hits from their batteries. Instead of shooting at random against the machine's bulk, now they could try to shoot at a few well defined points. "Provided that", as the Hunter said at the end of the meeting, "all replicators were the same, at least in their general architecture."

Campos was particularly enthusiastic about what they were discovering, even if he was not saying anything about the conclusions that he was starting to draw. Mike was aware that he was formulating a new theory about replicators, but he didn't pester him with questions. It was obvious that the scientific officer wanted to collect more evidence before telling them his conclusions.

They sent a few messages to the colony on Ceres to warn them of the danger and above all to suggest to them that they should not to broadcast any message that could betray their position. It was possible that the replicator had only an approximate knowledge of their position. If it had to waste some time searching for Ceres, they might be able to reach the colony in time to defend it.

These illusions dissolved when they got a request for help from the colony. The replicator had attacked when they were still twenty hours away. It was clear that the colony would be destroyed before they could do anything about it. They continued at top speed and, when they were four hours away, they detected the replicator. It was still in orbit around the third planet's moon and, from what could be deduced from the energy bursts coming from the surface, it was continuing with its bombing.

It was a few hours since the colony had given any signs of life. They prepared for battle, but were still two hours from their target when they realized that the replicator was leaving orbit. "It looks as if its work of destruction has been completed. Plot its course as accurately as possible: we need to find it after we leave Ceres", the Hunter ordered.

"Sir, we would do better to deal with the replicator first, and postpone any attempt at finding survivors. I don't think there are survivors, anyway, and if we delay our pursuit we might lose track of it", Romero pointed out.

"We cannot say whether there are survivors without attempting to find them. Above all I hope that some have been able to find sanctuary in those natural caves they showed us when we were on Ceres. If they followed our suggestions, they might survive for weeks there. Anyway, we cannot abandon them", the Hunter stated in a tone that brooked no disagreement.

It took about ten minutes to plot the course of replicator. "It is going straight toward New Shanghai", Bhagwat announced.

At these words Mike couldn't refrain from saying. "We cannot stop now, Sir. If we go to Ceres we cannot prevent the station from being destroyed. We must stop it immediately". He had always known that New Shanghai was in danger, but now that the danger had materialized he saw things rather differently. He thought of his friends and then of Ann. He had been able to keep her out of his mind until that moment, but now she suddenly resurfaced in his thoughts. And he realized how much he was missing her. He had to do something to stop that machine.

He made a more realistic evaluation of the situation. Even if she and Ms. Donovan had decided to follow his suggestion and get on board the first ship travelling towards safer areas of the colonized zone, it was impossible for them to have already left the station. Now that news of replicators infesting the system had started to spread, fewer ships would be docking at New Shanghai, and it might take months before being able to get one.

The Hunter's voice roused him from these thoughts. "How long will it take to reach it? And, above all, how long can we stop at Ceres to be sure we can get it before it reaches the station?", he asked Bhagwat.

The computations took a few minutes. "If we don't stop, we will get it in six hours, but if we stop more than forty minutes at Ceres we won't reach it in time", Bhagwat answered.

The Hunter took a quick decision: "So we must attack it right now. If there are survivors on the colony, they must wait until we get back: we cannot loose hundreds of people trying to save survivors that we don't even know exist".

Mike saw that Romero was indicating his approval of what he had said. He wanted to tell him that he had said it, thinking of Ann and the others, and that he was not particularly bothered about the reward, but said nothing. He realized that few on board would understand such a feeling. After all they were all hunters and they were there mainly for the rewards. From that point of view things had gone well in the last month. They had got two rewards and a third one seemed to be within their grasp. They didn't really bother that thirty thousand people had probably died on Ceres. They had actually saved them once and the situation now was not their fault.

They changed their course slightly to keep at a certain distance from the third planet and its moon, and they continued their pursuit at top speed. The replicator gave no sign of having seen them, but it was probably aware of their presence. It knew that they would go on chasing it and that, if it kept its course, it had a good prospect of destroying both the space station and the ship. Apparently replicators were only interested in destroying as many organic lifeforms as possible.

"Be careful and get ready", the Hunter told them, "it knows we are here. It will let us get close and then it will suddenly turn to hit us. When it does so, for a few moments it will divert all its available power to its engines and its shields will be weakened. That will be the best moment to strike". Then he turned toward Mike and asked him which one of the targets they had identified was the easiest.

"If this replicator is similar to the other one, we will have a good chance of hitting what we believe is its central control system, while it is turning round", he answered, showing a simulation of the maneuver on the main screen.

A red mark was superimposed on the image to show the exact spot to hit. Campos indicated his agreement. Clearly he had reached the same conclusion.

"Did you get that, Mr. Beaumont?", the Hunter asked. "If you can hit that spot while the shields are weakened, we can neutralize it with a single shot."

"In theory that's easy", Beaumont answered. "In fact, our weapons are not designed for precision shots. The weapons that have always been used for hunting are wide-field weapons, since nobody had any idea about which were the best spots to hit".

"Try to focus the beam better, and keep our aim fixed on that point", the Hunter commanded. Then he ordered everyone to put on their space suits.

Time seemed to have frozen. Everybody kept their eyes fixed on the screen, which showed that the distance was slowly being reduced.

Now Mike couldn't help thinking of Ann. He was feeling guilty. If he had been able to convince her to come on the ship with him, she would not be in the replicator's sights. Perhaps if he had insisted a bit longer …

His thoughts were interrupted by a sudden movement on the screen: the replicator was turning. "Too early. Hell, it is too early" the Hunter said in surprise. The replicator was still too far away and the front batteries were not powerful enough to place effective hits from that distance. The laser beams were aimed towards what they believed to be the central control system, but they were stopped by the shields.

"Ten torpedoes out, and get ready to fire as soon as the torpedoes hit the shields", the Hunter ordered. Now the replicator was closing fast. Suddenly a number of plasma-bolts were aimed at their ship and started hitting its shields.

"It is aiming at our engines" Mike warned them. "The shields are overloaded, they cannot hold much longer". Indeed it was not long before the shields gave way and the ship took a hit dead center on the engine room.

All the lights on the bridge went off, and it was filled with thick and irritating smoke. The restraint systems failed as well, and Mike flew toward the console in front of him. He didn't lose consciousness and he realized that he had been lucky. The console was close in front of him and he had hit in a fairly soft spot.

In the darkness he managed to find the controls for the emergency systems and was able to switch on the emergency lights and the backup life-support system. "Emergency life-support systems working", he announced. Then he checked the engine instrumentation and added "All propulsion completely lost".

He heard Campos' voice saying "It's leaving. It looks as if it is again on course towards the space station".

"It knows we cannot maneuver and that we will continue to drift on the same course. It can blow the space station up and then take its time in finishing us", Bhagwat added quietly.

Mike got up and realized he was stumbling. He walked toward the airlock, saying "I will go and see what condition the engine room is in. From here, I cannot tell whether it has completely gone, or whether we can restart the engines".

He sealed his space suit and went out through the closest airlock. When outside he moved towards the engine room, but soon reached a point where the hull just ended. From that point on the ship had just ceased to exist. "We have completely lost the engine room and the engines", he said aloud over his suit's radio system.

There was no answer. He looked around. He could see the replicator, which was slowly moving away, resuming its course and realized that now they couldn't do anything. They couldn't pursue and destroy it with their ship in such a condition.

He concentrated on his task, trying not to think of Ann. Working from the outside, he checked the airtight doors, which had closed properly and prevented what remained of the ship from depressurizing. There was not much that could be said about the situation: the replicator's aim had been quite good and, with just a few hits, it had disabled their shields as well as their propulsion system. And now they were no hope of confronting the machine when it came to destroy them after having destroyed the space station.

For a moment he fell prey to despair, but then he reacted. Perhaps they would still be able to evacuate the station in time, if he could warn them. Before leaving he had left all the instructions of how to modify their emergency capsules to reduce their cross section so that the replicator's sensors would fail to detect them. And Steve had promised him that he would take Ann well away from the station.

Floating in space, he moved towards the bridge, and when he entered the airlock he had come up with a plan.

20

Mike reached the bridge and looked in some disbelief at the scene in front of him. The smoke had thinned out and the emergency lights allowed him to see, in their ghostly light, what was happening. Some of the consoles had been displaced and were disconnected, and some of the screens had been shattered. Not many of the officers were still at their stations and Kimura was bandaging one of the crew, who clearly had a broken arm. He looked more closely and saw that many of the crew had hastily made bandages and were still covered in blood.

He realized that he had been lucky. He sat down at the transmission console, which was empty. After tuning to the New Shanghai wavelength, he warned them of the impending arrival of the replicator. "Leave the station immediately and scatter in space. Try not to betray your presence. Try to stay out as long as you can; and we will try to keep that replicator away from you", he finished.

"Congratulations. That was a nice speech", commented Romero, who had obviously heard him while he was speaking. "Can you please explain to me how we can do anything, now that we are drifting in space without shields?" Mike turned and realized that Romero was sitting at the Captain's station.

"Where is the Captain?", he asked him, without answering his question. "I must speak to him immediately".

"He is in quite a bad condition, Kimura answered. The impact hurled him against the control panel and he has some broken ribs. I fear the he may also have internal bleeding and I don't think he is in any condition to listen to you—provided he is still alive. Anyway, we have carried him to his cabin".

Those words impressed him even more than the blows from the plasma jets. He had to be alive, and Mike had to speak to him. He ran from the bridge heading for the captain's cabin.

He stopped at the door, which was half open, and forced himself to go in. The Hunter was lying on his bed with his eyes closed, and gave no sign of realizing that Mike had come in. Mike noticed that now he looked an old man. The energy for which he had been noted seemed to have gone.

"Captain …", Mike said to him quietly.

"What is it?", the Hunter asked, half opening his eyes.

"I must talk with you, Sir", Mike said.

"Mr Romero is now in charge. Refer anything to him"

"Captain, this is too important. I must speak with you".

"All right, but be quick", the Hunter answered, starting to cough.

"I believe we still have a chance, Sir. Or at least, we must still make an effort", Mike said, trying to overcome his own doubts. Previously, when he was in the airlock, he had thought that his plan was realistic, but now he started to have doubts.

"Go on", the captain said in a tired voice. But he tried to sit up in bed, lying against his pillows, to see Mike better. The movement caused another fit of coughing.

"Sir, I have a plan that could work", Mike went on as soon as he was able to listen. After a short pause, he said: "As you remember, I have that RG in my cabin. RGs, like all robots, can work in the vacuum of space without any problem. Now I can re-program it and strap the warhead of a small torpedo on it. Then I would go out on a shuttle and approach the replicator. When it starts to pursue me, I can retreat and release the robot at a very low velocity, so that the relative velocity between the replicator and the robot is low enough to allow it to go through the shields". He looked at the Hunter, who had not moved and seemed to be asleep.

Mike carried on: "The sensors of those machines are apparently set to detect organic life and any object fast enough to represent a danger. The robot should get through the shields and reach the generators. When it reaches that point I will ignite the nuclear warhead, which will cause a complete failure of its shields. I can then launch the second of the shuttle's torpedoes against its central control system, destroying it completely".

When Mike had finished speaking, they both remained silent for a while. Then the Hunter took hold of Mike's arm with his hand. "Mike, you are crazy. I appreciate your courage, but this is suicide. The monster will kill you even before you can get close enough. No, I cannot allow you to do such a thing".

Mike noticed that the captain had called him Mike. "Sir, it won't necessarily be that way. It is true that my plan is dangerous, but doing nothing is even worse. If we don't try something, we are all dead. First those on New Shanghai, and then us. That monster, as you have just called it, will kill us all, if we don't stop it. I'm not really risking anything, or rather, I am just risking dying a few hours earlier. But if I succeed …", he paused briefly. "If I succeed", he continued, "not only we will survive, but we will have tried out a strategy that could be used against other replicators, in other places. What is at stake is not only the lives of those on the station and our own lives, but something even more important. Please, let me try."

The Hunter remained silent for a while, which to Mike seemed like eternity.

"Perhaps you are right. Perhaps I should let you try. But Romero is now in command, and he won't allow you to do it."

"I can resign. Once I am no longer one of the crew, he cannot stop me from doing what I want".

"That won't work. You need one of the shuttles and some torpedoes. However we can sell them to you." The Hunter was thinking aloud. "At that point that replicator will be all yours. If you succeed, not only you will save your life or, rather, you will save us all, but you will become incredibly rich … Well, if you want to try, go on, do it", he concluded.

Mike thought he was dreaming.

"Well, but when you go to get on that shuttle everyone on board will realize that you had an RG with you …", the Hunter went on.

"No, Sir. There won't be an RG to be seen. What they will see will be an RW, the first RW to be used against a replicator", Mike answered.

"RW?", the Hunter asked.

"Of course, RW. For Robotic Weapon, that is".

"Well, I will tell Romero to sell you what you need. Good hunting, Mike", the Hunter concluded, falling back on his bed.

21

Mike turned and saw that Campos was in the room, near the door.

When he was close to him, the scientific officer said, in a low voice: "So you had a RG on board! Now I understand why you have remained hours in your cabin without showing up."

Mike started walking fast towards his cabin, followed by Campos. "I was trying to get beyond the usual limits of artificial intelligence. I was certain that I would be able to transform it into a true thinking being".

"Clearly without having much success. Don't think you are the first to make such an attempt. But now let's think about more serious matters: your attempt will never work", Campos continued.

Mike suddenly stopped. "What do you mean, Takashi? I know that it is a difficult and risky business, but I hope to strike the replicator's central control system with a torpedo, after the robot has caused a total shield failure."

"Well, let's assume that you succeed and its control system suffers serious damage. Do you understand what will happen then? A replicator is built to replicate itself, by definition. I dare say that immediately after your nuclear warheads have damaged the shield generators and the control system, the self-repair systems will start working and within a few hours the machine will be in perfectly good health—if the term health can be applied for a object of that sort. To destroy such a monster you need something much more powerful than a few small torpedoes like those you can carry on a shuttle."

"But then there is nothing we can do to save those on New Shanghai", Mike answered. He was really disappointed.

"That's not what I said. If I come with you and you are able to blow up the shield generators, I think I will be able to take control of the monster and neutralize it for good", Campos said.

"But how is that possible, if you say that it is able to repair any damage that we can inflict on it?", Mike asked.

"It's just a theory, but I think it is worth a try. We can in any case carry some more torpedoes with us, provided they can fit in the shuttle and, if I am wrong, we can try to damage it more seriously so that it is out of combat for a longer time".

Seeing Mike's upset expression, he added: "Don't worry, I don't want to take all the credit and all the reward. The idea of attacking the replicator with a shuttle is yours, and if you buy the shuttle, you will be the owner and the captain of the ship. I will settle for the title of first mate … Anyway, I am not able to fly a shuttle." Then he added: "Before we begin, you should tell me where you got the idea of using a shuttle to attack a replicator."

"Well, first of all, a shuttle is all we have left. And then Ann, that girl on the station …"

"Sure I know her, I saw her with the RGs while you were playing out that comedy with the waiter", Campos interrupted him.

Mike looked at him suspiciously. Campos realized why and added: "Don't worry, I am not like Romero … Go on, what has Ann to do with shuttles?"

"She kept speaking of a nineteenth-century book, describing whaling. Have you read it?"

"You mean Moby Dick? Yes, at school. And, like all the books I read at school, I remember little of it", Campos answered.

"Well, I found a copy of a movie three hundred years old and I watched it ... In the movie, they launched small rowing boats from the main ship to get close to those beasts. I don't know why, perhaps their ship was not manoeuvrable enough, or perhaps there was some other reason. I don't know. Those scenes were impressive and I think they influenced my idea of using a shuttle to get close to a replicator", Mike went on.

"But now let's go ... while we carry on talking, our replicator will will be getting farther away. I'll go to my lab to prepare what I need, and you should start modifying that RG of yours. It's a pity replicators are not like Romero: it would be easy to get that whore to convince it to lower its shields", the scientific officer concluded, walking away towards his lab.

Mike ran to his cabin. As soon as he entered he switched on his computer and then switched on Lulu as well.

While both machines were bootstrapping, Mike sat on his bed and opened the box where he kept the tools he used to work on robots.

"Lulu, undress and come here", he ordered.

Lulu came near and, starting a dance, asked: "Do you want me to put on some music?"

Mike realized that such a request would simply trigger that behavior. "Stop, I just told you to take your dress off", he said harshly.

Lulu took off everything she had on and knelt in front of him with a submissive expression. "Do you want to punish me, master?" she said in a very low voice, remaining motionless.

Mike remained motionless too, surprised by that reaction. Then he understood: the tone of his voice had called from the depth of her memories a program he had not suspected to exist. 'But a generic RG should not have programs of this kind' he thought. Then he realized that someone, perhaps in that night club on Helium-mine, must have downloaded some programs and added them to Lulu's standard programming. He knew of the existence of software packages to introduce such reactions in the behavior of RGs, but he was surprised he hadn't realized it while working on the robot. 'God knows what surprises you might still have in store for me ...' he thought. But now it was too late to find out.

He realized that he was wasting precious time with such thoughts. He stretched out his hand and noticed that with hardly noticeable movements Lulu had tilted her face to the right and shifted her knees. If he were to slap her face, she would hit the floor.

Mike couldn't help noticing that, whoever had tampered with her programming, must have been a professional. He put one of the fingers of his right hand between her shoulder blades and, pushing hard, found the emergency switch.

As soon as he pushed the switch, Lulu froze.

"Farewell, Lulu", Mike said in a low voice. "You were just a robot, but nevertheless I liked you. I will never know whether you could have become something more …"

He took a tool from his toolbox, and started removing the layer of synthetic flesh and skin that covered the electro-mechanical parts of the robot. He was trying not to think to what he was doing. He regretted having to destroy that body, although he knew very well it was little more than a statue or a doll.

As soon as the head was uncovered, he connected a cable to the socket located on the back of the first vertebra, and started to control the robot directly from his computer.

He completed that work in less than a quarter of an hour and by then he had in front of him just the metallic skeleton of a robot, with all the actuators, the electric motors, the pumps and the circuitry completely exposed. However, it was quite upsetting, because of its similarity to a human skeleton. On the floor there was a sort of a distorted copy of a human body, a sheath emptied of all its content. He had at least taken care to put her face downwards, so that he couldn't see the eyes, which were now empty even of the expression that tried to imitate a human expression.

He took some measurements of the skeleton in the belly area, to check whether he could fit the nuclear warhead from a small torpedo in there. There was not much space, but it should be enough.

Then he started to re-program the robot. He erased Lulu's personality, without touching the low-level controls, and downloaded many guidance and navigation programs onto the robot's computer. He was surprised to see how large a memory had been necessary in a robot built to emulate human behavior. After downloading all the starship's navigation and guidance programs, he also loaded those of the shuttles and of the mobile maneuvering units, only to realize that its memory was still half empty.

That was clearly an important feature. He could insert many more instructions, which might perhaps be important for the success of his plan. He called Campos on his personal intercom. "I am ready. Can we go?"

"I have still a lot of things to do, but I can carry on with that while on the shuttle", Campos answered.

"Well, we can meet on the bridge", Mike concluded, closing the channel.

He checked everything once more, then he got up, took the tools he thought might still be useful, and started walking toward the door. "RW, follow me", he ordered.

22

When Mike entered the bridge everybody turned to look at him. The situation had slightly improved, at least now they were all at their stations, even if many sported bandages, but nobody looked as if they were doing anything useful. Actually there was little to do on a ship adrift in space, with propulsion out of action and the generators at minimum power. The shields had barely enough power to protect the hull against the impact of micrometeorites, while the ship was continuing to follow its former trajectory. He was still looking around when Campos also came in, pushing a trolley carrying two large crates.

"The Captain told me that you had resigned and that you plan to leave us to attempt an attack against that monster", Romero told Mike, looking at the strange robot that entered the bridge after him. "And that thing, what is it?", he added.

"Yes, sir, and I need to buy one of the shuttles, with its standard equipment, and that also includes the torpedoes", Mike answered, ignoring the question.

"Yes, we can do it. I warn you, however, that it will not be cheap", Romero said.

"How expensive?", Mike asked. After all he was not really interested in the price. If he succeeded he wouldn't have any problems with money, because the reward would be almost all his, whereas if he failed the problem would only matter to his heirs—if he had any. At present he was only interested in not wasting time.

"Half a billion, not a credit less", was the answer.

Mike was shocked. "What? With that money I could buy a starship …"

"If you don't agree, you can buy what you want somewhere else. Here and now that's the price".

Mike understood, and realized that there was nothing that he could do. He signed the contract on the ship's computer. "Anyway, if I fail, that replicator will kill all of us, and nobody will cash that money. You had better keep your fingers crossed all the time". Then he spoke to the robot: "RW1, go onto the shuttle and prepare to leave".

"Yes, Sir", the robot answered, military style.

Mike was satisfied. He knew that Romero would suspect that the robot was RG46, but when he had modified the program he had thought that "yes darling", said with Lulu's sexy voice was not an appropriate answer for a RW. If he had more time, he would have programmed a military salute, with its right hand rising to the side of its head, but he had had no time to bother with such details.

"What does that mean: RW?", Romero was unable to refrain from asking.

"Robotic Weapon. And you will soon see it in action", Mike answered as he left. 'And pray it actually works' he thought. Campos also said goodbye to Romero and followed Mike out.

When they reached the shuttle, the robot was already sitting in the co-pilot's seat.

"Let's go. Set course toward the replicator. Top speed", Mike ordered, sitting at the pilot's station and activating his restraint system.

The robot flew the shuttle out of the ship's hangar and started accelerating toward the replicator. To Mike's satisfaction, the manoeuvre was perfect. "I shall go and get the things I have brought with me ready", the scientific officer said. He was amazed by the fact that a robot that was just a modified RG, could fly a shuttle.

"I have a lot of things to do to get RW1 ready for its encounter with the replicator", Mike replied. "I'll see you when I have finished". He ended by saying "I have a lot of questions for you".

Campos went directly to the cargo bay, opened his crates and started pulling out the contents.

When they were in the open space, on course toward their target, Mike got up and went to the weapons bay. He started dismantling one of the torpedoes and very carefully took out its nuclear warhead. He nestled it into the lower part of the robot's body and checked that all the connections to the weapon's ignition systems were fully in order. Then he took two individual mobility systems and attached them to its shoulders.

The work required more than four hours, but at the end of it the robotic weapon, which he hoped would disable the replicator, was ready. He ordered the robot to enter one of the launch tubes and went back to the cockpit. He quickly ate some emergency rations and went on with his work. He now had to finish programming the RW. He went on until was so sleepy he could no longer concentrate on what he was doing.

He got up and went to the cargo bay, where Campos was still working. "How is your work on that robot going?" Campos asked him without taking his eyes off what he was doing.

"I am getting there, but things are more complicated than I expected", Mike answered. "I need to be sure that it can approach the replicator slowly enough not to be stopped by the shields. And that it could identify the generators. The problem is that I have to program it from scratch".

"I guess that the programs of RGs are not of much use for that", Campos commented, again starting to fiddle with some of his electronic components.

Mike went closer so that he could see better: Campos had set up one of the crates so that he could use it like a workbench, and was connecting a tangle of electrical cables to a small computer, which, in turn, was connected to the shuttle's computer.

"Can you please explain to me what are you trying to do? If it is true that we cannot destroy that replicator with the torpedoes, the lives of all of us are entrusted to that contraption you are trying to put together. I daresay it is not a pleasant thought, at least judging from the aspect of that whole tangle of wires".

Campos stopped working and turned toward him. "No, you are not wrong, but you will see: it will work. But now I have to stop and get some sleep, if I don't want to make any mistakes", he ended, getting up and walking toward the cockpit.

They settled on the cockpit's seats, the most comfortable places on a shuttle for anyone to get some rest.

"I will reduce the artificial gravity, so that we are more comfortable. Be careful if you get up", Mike said pushing some buttons on the control panel.

"Where did you learn to fly these shuttles? I didn't think it was part of a maintenance technician's job", Campos asked.

"I told you that my family owned a small cargo ship. I learned to fly when I was a boy, with my father. You know, you grow up quickly on those ships. Everybody needs to work, and as soon as my father thought I could do it, I had the task of flying the shuttles", he answered, lowering the lights and getting ready to sleep.

He was so tired he couldn't sleep. He was wondering what the scientific officer had in mind. He had spoken of taking control of the replicator, but he had no idea of what those words meant.

After a quarter of an hour he said very quietly "Are you asleep, Takashi?".

"No, I cannot sleep. I keep thinking of all the work I still have to do and to the little time we have", he answered.

"But what is your plan? What did you mean when you said we must take control of that replicator?"

"Right, I'll explain. In any case, I cannot sleep. I read all the documentation I could find on what they thought about Von Neumann machines at the beginning of the space age." He stopped for a moment, and then asked "You know what Von Neumann machines are?"

"Of course, I too read a lot about them. And every time I think about it I wonder how they imagined they could use those machines to explore space".

"You see, it all comes from relativity theory. The interpretation they had at that time implied that you cannot travel faster than light", Campos started. "Strictly speaking, that is true, but we can distort space-time using the warp drive …" Mike interrupted him.

"A theory of warp drive was developed at the end of the twentieth century, but in a form that wouldn't work", Campos went on, ignoring his interruption." And at that time they had no energy source powerful enough to reach high speeds. They had strong reservations against using nuclear power in space …"

"But that's absurd! How could they think of travelling in space without nuclear energy?", Mike said.

"Clearly they were not able to do very much in space. But don't keep interrupting me like that!" Campos was getting nervous. "They needed decades to get used to the idea that only nuclear energy would allow them to build an advanced civilization, and above all a spacefaring civilization. Anyway, many thought that humankind would never be able to leave the Solar System, and that only robots could explore interstellar space. Von Neumann machines appeared to be the only solution to the problem, above all since they had not yet understood the limitations of artificial intelligence. They were doing like you, when you tried to transform that RG into a girl, or else like Father Dubois …"

"I didn't ask you for a history lecture" Mike interrupted again. "I just wanted to know how we can take control of that robot after RW1 has destroyed the shield generators".

"Just a moment, I'll get to that", Campos answered in a reassuring tone. He clearly realized that he had gone too far with his history lecture, but he was genuinely interested in the history of the last centuries.

"I think that the civilization that built the replicators had an approach that was not very different from that of our ancestors. How do you explain the fact that those machines, which are so advanced in many respects, cannot travel faster than light? It doesn't make any sense that whoever built them didn't make use of the most advanced technology available to them. The only explanation is that they were originally probes for interstellar exploration, like the Von Neumann probes that Tipler suggested."

"But it makes no sense! Why program a probe to destroy organic life?", Mike countered. "They are not probes, they are weapons".

"Wait. Some time ago I found an old paper from the end of the twentieth century where it was stated that the strategy of using Von Neumann probes couldn't work since, once they were far from their builder, they would start mutating and evolving, following a more or less Darwinian scheme. Well, replicators were not designed to destroy organic life. They somehow evolved in that direction. At any rate, whatever happened, it is reasonable to assume that the builders of those machines had built their control system in such a way as to take control manually. All what we need to do is to find the correct interface."

Mike was appalled. "Takashi, the Hunter told me I was crazy to try to attack a replicator with a shuttle. But you are infinitely more crazy than me. Do you realize that all our hopes rest on the possibility that, in a few minutes, at best in a few hours, we get into that replicator, we find that sort of back door you assume, and like hackers, we get into a computer about which we know nothing, and take control. We will never succeed".

"Give me time; what I told you is a theory that I have been formulating for years, and I would never try to do something like that. However, recently I got some evidence. Two of the fragments we obtained confirm what I said. First, we got a fragment of a hull thick enough to maintain a pressure suitable for organic life. Replicators include a zone, perhaps a small one, where their builders could live".

"I never heard anything like that", Mike said. "Replicators have no hull, they have an open structure and have no interior. All the information we have obtained confirms that".

"I didn't say they have an air-tight hull like space ships. I said that there must be a pressurized zone that can be inhabited by organic beings like ourselves. Nobody would build such a thick structure unless it were to create a pressurized zone. But the other fragment is even more interesting. It is from something that cannot be anything else than a keyboard, not much different from those used with our computers".

"But that would imply, for one thing, that the hypothetical builders of the replicators had fingers ... That is too anthropomorphic to be believable. And then, that they communicated with their computers exactly in the same way as we do with ours, which is another anthropomorphic statement. Are you sure of what you said?". Mike's voice was both thoughtful and disbelieving.

"I know, it seems to be too anthropomorphic to be believable. But after all, any technological species must have organs to manipulate objects. And any manipulatory organ must have fingers of some sort. From the size of the keys it looks as if their fingers were larger than ours. At any rate, the second point is not so strange. A keyboard of some type is the ideal instrument to communicate with computers, at least with the simplest ones, and anyone who built a computer must have used a keyboard, at least initially. We still sometimes resort to keyboards, even now that we mostly use voice control to communicate with our computers."

"Of course we use a keyboard. That way you can control a computer even before all the software for voice control has been uploaded", Mike said, thoughtfully. "When I transformed RG 46 into a weapon, I had to use a keyboard to control it after cancelling all the high-level functions and before uploading the new ones!"

"Well, so replicators may well have devices of that kind if their builders wanted to be able to take control in any time, to modify their programs."

"And so we get on board that machine, we get to the control room, we take the keyboard and, by modifying its programs, we turn that monster into a lamb. No, it's too easy, it cannot work!" Mike exclaimed.

"No, it's not easy at all. As I told you before, replicators evolve and so it cannot be taken for granted that, after so many generations, the manual controls, which were not used for thousands and thousands of years, can still be used as when they were designed". As he was speaking, Campos turned towards the control panel and saw the clock. "My God, do you see what time it is? We must sleep, tomorrow we still have such a lot of work to do".

"You are right.", Mike answered, stretching out on his seat. Now he was much less nervous, and fell asleep right away.

23

The following day they started their work as soon as they woke up. Mike was working quickly, and from time to time he had the feeling that he would be unable to finish in time. Because of the way he was working, he was unable to think of anything else. The space station, the Hunter, Ann, his friends, all had ceased to exist for him, all his thought were focussed on their shuttle, RW1 and the replicator.

Suddenly, late in the morning of the third day, he realized he had done everything that was to be done. The robot was ready, together with everything else. Obviously he was not sure that everything would work as he hoped, but he had done all that he could do.

He closed his eyes and fought against his desire to start again from the beginning, checking everything once more. He got up and went along to the cargo bay, where Campos was still working at his computer. He was about to start to speak, but the scientific officer gestured to him to keep quiet. Clearly he still had work to do.

He went back to the cockpit and again switched on the passive sensors to try to identify the replicator. It took him some time, but then he found it against the star background, more-or-less where he thought it had to be. On the other hand he couldn't find the space station. It was still too far away, but he knew that it had to be in its proper location and, in any case, there was no point in attracting the attention of the replicator by switching on the active sensors.

He computed the trajectories again only to realize that they couldn't reach the replicator before it closed in on the station. Even if they succeeded in their scheme, there was nothing that could be done for New Shanghai.

There was one last attempt he could make. He switched on the transmitter at full power and keyed in Joe's address.

"Joe, there is a replicator aiming directly at you. It is impossible to stop it. Leave the station and scatter in space and, above all, keep a strict radio silence".

That transmission had two goals: the first, the obvious one, was to warn the station, so that they could bail out using every available spacecraft. The second, less obvious one, was to lure the replicator towards them.

He was ready to change the shuttle's course in case the replicator decided to postpone its attack on the station to deal with them first. From what he knew about replicators, Mike realized that its choice depended only on how certain the machine was about the position of the station and the number of people on it. If it was sure that the station contained a large number of organic life-forms, it would probably not change its course to attack what it had certainly identified as a shuttle, which would have only a few people on board.

He waited for half an hour, and when he was sure that the replicator hadn't changed its course, he repeated the message using even more power.

The replicator gave no sign of attacking them.

Mike tried to evaluate how much time they had before the battle. Within ten hours the replicator would be in range of the station, and the shuttle would reach the same location two hours later. If the survivors could remain hidden in space for two hours, they could make it. The situation was not desperate, provided they could scatter widely enough.

He had to sleep. He stretched out on his seat and switched on the autopilot. Then he took one of Joe's pills, set its timer for seven hours and swallowed it. Almost immediately he was asleep, in a heavy and dreamless sleep.

The first thing he realized when he woke up was that he had only a moderate headache. 'It looks like I am getting used to these things', he thought.

He turned his head and saw Campos in the other seat.

"Good morning Mike. I have also finished my work. Now we have only to wait, hoping everything will go well".

Mike looked at the screens. The replicator was there, in front of them, dreadfully close, and it was possible to see the space station in the distance. Thinking about it, he realized that the replicator was not that close, but only terribly large.

Mike started to panic as he looked at the screen. Some sequences from that old movie, Moby Dick, that he had found in the entertainment area of his ship's computer, came back to mind. When he saw it for the first time he had wondered how was it possible that those ancient sailors had dared get so close to those enormous beasts in such small boats, and perhaps this had suggested to him that he could use a shuttle for his own mission. But now he was really on board a small vessel, at close quarters to an even more dreadful monster.

He felt he was the equivalent not of Ismael, but of Ahab himself. 'I am as crazy as him, and I deserve the same sort of death', he thought. But it had got to the point where he could do nothing else than go on. If that was his destiny, now he had to go on with his personal war against the monster, exactly like Ahab.

Then suddenly his mood lifted. 'At least I don't have to row, and I won't get splashes of ice cold water on my face', he concluded, trying to free himself from such gloomy thoughts.

To distract himself, he checked all instruments again. Then he broadcast more messages, trying to lure the replicator and distract it from its attack on the station. But it continued its course straight toward New Shanghai.

Both the replicator and the station were now much larger on the screen. No message, light signals or other signs of activity came from New Shanghai. Mike wondered whether what he was seeing was now just an empty shell or whether there were still people inside.

He hoped that Ann were somewhere in space in front of him, on a well hidden emergency capsule. If so, she had some hope of surviving.

He called Campos, who was again in the cargo bay, and told him to get ready.

"Now it's up to you. I am checking everything again, and am getting ready for the next phase", Campos replied.

Suddenly a number of bright plasma jets shot out from the replicator. When they hit the station it blew up. The scene lasted only a split second. One moment the station was there in front of him and the next it had simply disappeared. Mike silently prayed that there had been no one on board.

After the attack, the replicator slowed down and started to move to one side, as if it was looking for something.

24

The replicator continued combing space around the position where the station had been until a few minutes earlier. It was clear that it couldn't understand where all the organic lifeforms which it assumed were on board could be, because it was unable to identify what should have been their remains. That was the moment when Mike could attract its attention. He switched on his transmitter again and started broadcasting at full power.

The replicator clearly was aware of his presence, since it turned toward the shuttle and for a short time remained hanging there, apparently uncertain about what to do. It had clearly realized that there must have been many organic lifeforms in space around it, because it was impossible for the station to be empty or for all of them to have suddenly vanish. It also realized that there was a shuttle with at least one form of life, in a well-determined nearby location. And whereas the former was just a hypothesis, the latter fact was definite.

It finally took a decision and started to accelerate towards the new target. Mike turned the shuttle and started to accelerate directly away from it. "This is it, Takashi. It's coming, I am ready to launch RW1", Mike said using the shuttle intercom.

"Well, Mike, go ahead as planned. I am ready too", Campos answered from the cargo bay.

When the shuttle was at maximum speed, he launched RW1, using a burst of compressed air, and then started to observe it with the telescope. "Farewell, Lulu", he said quietly, realizing that in fact Lulu had ceased to exist a few days ago. What he had launched was just a weapon and nothing more. However he couldn't help thinking of that skeleton when it was covered by synthetic flesh and skin.

By now the shuttle was fleeing from the replicator, which was following it at its top speed. In between them there was the robot, which was maneuvering using the cold gas jets from the maneuvering units and was slowly closing on the replicator.

Mike kept his telescope aimed at the robot and adjusted the trajectory of the shuttle to keep the robot between the spacecraft and the replicator. As planned, RW1 was in a horizontal position, with its head toward the shuttle and its feet toward the replicator, such as to expose as small an area as possible to the latter.

After about a quarter of an hour the robot was within range of the replicator, but the latter didn't take any action. It was obvious that it wasn't aware of its presence.

Mike continued to follow the scene through the telescope. After another half an hour he realized that the robot should have arrived at the shields. He had explained to Campos that the shields shouldn't be able to stop an object that approached at a very low speed, but this was just a theory, based on how the shields on terrestrial spacecraft behaved. Nobody could say anything about how the shields of the alien machines worked.

As the robot continued its attack run, the replicator showed no reaction whatsoever. Mike left the telescope for a moment to check the distances shown on the screen. He realized that things were not going as well as they appeared. The replicator was closing slightly faster than predicted. If RW1 didn't hurry up, they would be within range quite soon.

He went back to the telescope. Then he saw the robot stretch out its arms and take hold of one of the replicator's beams. It was now inside the volume protected by the shields and had reached the actual structure.

"Here we are, RW1 is on its target. Hold on, I shall soon start evasive maneuvers", he told Campos, who acknowledged the warning.

RW1 started to move along the truss, working its way inside the huge machine, but Mike was unable to watch any longer. He had to start maneuvers to fend off the shots from the replicator, which was now in range. It started to fire a plasma bolt, which hit the outer region of the shuttle's shields. Because of the distance it was a weak shot and didn't cause any damage, but the replicator was getting closer. 'Hurry up, RW1, or else everything will turn out to be in vain,' Mike thought as he gave a jerk to the controls, causing the shuttle to move suddenly sideways.

A plasma jet passed on one side. While he was preparing another evasive maneuver, he glanced at the telescope's screen and saw RW walking on one of the replicator's beams, making for the shield generator. Just at that instant the machine made an abrupt sideways movement, followed by other jerks to the right and left. It was now aware of the object that was moving inside its structure and was trying to shake it off.

'Hold on, RW1, the rodeo has started …', Mike thought.

The robot took hold of the beam from which it was hanging with both hands and pushed itself off hard, flying in the direction of the shield generator.

'Here we go', Mike thought, turning the shuttle so that the bulk of the spacecraft was in between him and the replicator.

A blinding light flashed on the screen. 'It's gone', Mike thought.

The replicator stopped firing and all movement ceased. Mike looked more closely and realized that RW1 had disappeared, vaporized by the small nuclear warhead together with the shield generator, but the rest of the body of the replicator looked undamaged. The fact that the machine was not moving, however, meant that the gamma ray burst had damaged the control systems. Mike looked at his instruments and was quite glad that he had turned the shuttle at the moment of the explosion: 'If I hadn't', he thought, 'that explosion would have fried me too'.

He told Campos that the replicator had been momentarily disabled.

"I'm ready. Everything I need has been packed and I am going to the airlock to put on my space suit", he replied.

They had no time to waste. They didn't know how long would it take for the self-repair systems, which the replicator must have, to fix the damaged control systems.

Mike flew the shuttle straight toward the replicator, hoping its shields were no longer operational, thanks to the destruction of the generators. He aimed the torpedoes that he still had towards what he believed to be the central control system and for a moment he was tempted to shoot. Even though he knew that he couldn't destroy it permanently with the weapons he had and that their actual plan was different, the temptation to go back to his original idea was very strong.

The shuttle raced towards the replicator, which was drifting in space with apparently neither propulsion nor control. It entered the space that should have been protected by the shields. A little later, it entered the machine's interior through one of the many gaps in the lattice structure. Now Mike was moving cautiously inside the replicator. The structure was extremely complex, surrounding a huge empty space, where the components of the new machine were assembled during the replication process. All around he could see arms and other robotic systems, which for now were inactive. Then Mike saw something moving. He looked more closely. One of the arms had been activated and was moving away from the main structure. The systems in charge of maintenance were starting the repairs to fix whatever had been damaged by the gamma ray burst.

Mike took aim with the shuttle's laser and fired. The arm was vaporized instantly. He didn't want to inflict any more damage on the replicator, but he couldn't allow the machine to start working again too quickly.

He had to hurry up. He aimed the shuttle directly toward the zone where he was sure the central control system was located.

Suddenly he realized he was there. In front of him there was a sphere, apparently made of metal, with a few small openings, connected to the rest of the structure by a few beams. A robotic arm, carried by a small modile platform, arrived from somewhere and attached itself to the sphere. The arm opened a small hatch and started working. It was clearly reactivating one of the control devices. Mike destroyed it with a single shot.

He flew slowly around the sphere until he saw an opening larger than any of the others. It was as large as a door, and suitable for a human being to enter. He moored the shuttle close to the door and went to the airlock. While he was donning his space suit he saw Campos, who already had his suit on and was carrying a big backpack full of instruments. A cable was sticking out from the backpack.

"We must hurry up. I don't know how long the maintenance system will take to restart the main computer", Campos said, sealing his suit. Mike did the same and they started depressurizing the airlock. Soon they were out in space, floating slowly toward the opening. The first thing Campos did was to connect the cable from his backpack to the shuttle's external data connector. Seeing Mike's intrigued expression, he explained that he had prepared an interface to connect the shuttle's computer directly to that of the replicator.

Mike was leading the way, while Campos followed, unrolling the cable. As soon as they reached the opening, Mike started examining it by the light of his inspection lamp. There was a door hinged inside the opening but, as Mike noticed immediately, the shape of the door and that of the opening, although similar, were not identical. The door could not be closed. Moreover, there were some remnants of a seal, but they didn't go fully around the door.

"This door cannot close", Mike commented.

"It looks like it", Campos replied. Undoubtedly, in the original design the door was made to close, but now, after so many generations, mutations had changed its shape and made it useless. He was elated by discovering that things were exactly as he had predicted. And if he was right so far, it was not so unlikely that he might be right on the other points as well.

Actually it was impossible for this door to have been damaged or worn out. The replicator was brand new, built just a few days before, and there was no sign of any damage around it. It had been built exactly like that, unable to work.

They went in. Inside the sphere there were a few control consoles and some seats. They looked at them carefully. They were similar, but not identical to one another and their shape was irregular. It was clear that originally they had been designed for some humanoid lifeform, but by now they would be quite uncomfortable even for their designers.

Suddenly Campos stopped. "Do you realize we are the first humans to see an alien artifact in detail?" He was overawed by the implications of the statement he had just made. Mike brought him back down to earth. "Let's start work if we don't want this alien artifact to turn us into little pieces".

One of the consoles was larger than the others and above it there was what could have been a screen. Actually it was just a more-or-less flat surface, with an irregular shape, some parts of which appeared to be provided with light emitting materials, although these were, in fact, unlit.

The whole thing looked like the control room of a spacecraft, drawn by a boy who had absolutely no idea of what the various parts were meant to do nor what their function was.

Something looking like a keyboard was lying on the console. Mike tried to press some keys, only to realize that the objects, which had the shape of keys, were just rectangular pads protruding from the surface. But suddenly one of those key-like objects yielded under the pressure of his finger and a light appeared on the screen. "It works!", Mike said, surprised and excited.

Campos put his backpack on the floor, where it started to float becasue of the lack of any artificial gravity. He opened one of the pockets and took out an instrument that produced a narrow laser beam, able to cut through any material. He started cutting the surface of the keyboard around the working key. There was a wire made of conducting material. He followed it and at a certain point he found something that looked like a connector. Fixed to the other part of the connector there was an electrical cable.

He went on cutting the console, following the cable. Now the cable looked like the real thing, with a number of insulated wires, made of metal, most probably gold, inside it.

"The designers used the best materials for their electronic devices. They didn't bother about the cost", Campos said.

"Gold is quite common on asteroids" Mike answered. "It's not surprising that they used it where it would be of use".

It was an interesting fact that this alien technology looked so similar to Earth's technology. These were actual cables, made in such a way that any human from Earth would understand.

As Campos was working, they felt a heavy blow against the outer surface of the sphere. Another of the robotic arms had started to repair the central control system.

"Shall I go out and stop it?", Mike asked.

"No, there would be no point", Campos answered. "They won't be able to reactivate the various systems in time to get control". Or, at least, he hoped so. He continued cutting the console until he found another connector. He studied it carefully: it looked fully operational. He disconnected it and carried on, making a note of all the connections between the wires.

The number of wires attached to the connector was not very great, and Campos started connecting them, one by one, to the interface he had in his backpack. Then he plugged it into his computer and started studying the situation. He had prepared a number of programs to analyze the communication protocols that the aliens had used to program their replicators, and his work was progressing quickly.

While he worked, he realized that, as the robots working outside the sphere repaired the damage caused by the nuclear explosion, the replicator's control system started to power up again. But each time a system returned to life, Campos was able to take control through his interface.

After about an hour they realized that one of the main thrusters had started working. Campos switched it off and then back on again. He had complete control of it.

The interface was not only able to control the replicator's various systems, but it was also able to receive the signals coming from the various parts of the machine. It was now clear to them that not only were the builders of the replicators organic beings with a humanoid body, as they could tell from the shape of the seats, but they had to be able to see, and to recognize shapes in ways that were not very different from those of humans.

On the screen of Campos' computer they could see, as they watched, an increasingly detailed drawing of the replicator was appearing. Each time Campos activated one of the parts of the machine, the corresponding part of the drawing lit up. The area where the warhead carried by RW1 had exploded was still completely dark. Then, all of a sudden, some lights started to shine in that area as well, a sign that the self-repair system had started to activate the shield generators as well. When that section was fully illuminated and the generators started working, Campos switched them off to check that that particular subsystem was also under his control. Then he switched it on again.

When he was sure that everything was working properly, he switched off the various systems, one by one, connected the end of the cable that he had carried from their shuttle to the replicator's computer and started moving toward the door.

"Is everything all right?", Mike asked him.

"I venture to say that it is. Now we can move to a pressurized area and take off these bloody space suits", Campos answered. He was not used to working while wearing a space suit and felt quite uncomfortable.

"What bothers me is rather the absence of gravity than working in a space suit", Mike replied. "By the way, why do you think they didn't install an artificial gravity system in here?"

"Maybe it was here in their original design, and then evolution had just eliminated it, because it is of no use to replicators. But perhaps it is more realistic to imagine that the replicators' builders had not developed the technology. If you think about it, artificial gravity is similar to warp drive, which they clearly didn't know about".

Slowly they returned to the shuttle and went on board. They had worked for more than ten hours outside and were quite tired. Particularly Campos, who was not used to extra-vehicular activity. They left their space suits in the airlock and Campos immediately switched on the first computer terminal he found. He could see the replicator's computer as a peripheral of the shuttle's computer. Not a peripheral like a printer or a generic input/output unit, but rather like a slave unit.

"It works perfectly", Campos said in a triumphant voice. He switched off the terminal and followed Mike who was heading towards the cockpit.

"After all that work, it is the least it could do", Mike answered, sitting down on the pilot's seat.

Campos sat down on the co-pilot's seat and again tried to contact the replicator's computer. He switched all the systems on and then off as if, after all that, he didn't really believe that his plan had succeeded and he was in control of that huge alien machine.

Mike realized he was exhausted, and that now he could rest. But he had still one more thing to do. He switched on his radio and called Joe. "Joe, tell everyone from the station that the emergency is over. The replicator has been neutralized. You can switch on your transmitters and set course toward the *Morning Star,* which is coming this way". He then broadcast the coordinates of the ship.

Suddenly several transmission sources appeared in the space ahead. He tuned in the frequency of the safety capsule that he had prepared with Joe and Steve, and his friend appeared on the screen. He was at the controls of the small spacecraft, but behind him he could see many people.

"Did you really succeed? How did you destroy that monster? We just saw a short flash, like a small nuclear warhead, not enough to destroy a replicator!"

He realized, from the short time the answer took to reach him, that the emergency capsule had to be no more than a million kilometers away.

"Don't worry, Joe. I didn't say that we have destroyed it. Campos and myself have taken control and it is no longer dangerous". He paused, and then asked what really mattered to him. "Who is with you? How many people were able to get out of the station in time?".

The few seconds needed for an answer to arrive seemed like eternity to him. Instead of replying, Joe panned the camera so that it showed all people on board. The safety capsule was carrying so many people—more than he had ever thought could fit in there. He saw Steve, Ann and Madame. Mike took a deep breath and was hardly listening when Joe answered "Almost everyone was able to escape. As soon as we got your message, all ships that were docked here left, overloaded with people. Then we launched all the emergency capsules: eight of them with an average of twelve people in each. There are about a hundred people here in space, waiting for someone to pick us up".

"Go to the Hunter's ship, as soon as I can I will join you there and I will tow you all to the nearest harbor". Then he went on: "Ann, I have to speak to you. See you on the ship". Then he closed the communication channel. He didn't want to speak to her in front of all those people. Moreover, the fact that he had succeeded in capturing a replicator was so exceptional that what he was saying would have been reported in every newspaper on Earth and the colonies. "There was Ms. Donovan as well, if I am right in what I saw", Campos said. "Do you know her?", Mike asked, surprised.

"When we were on New Shanghai, instead of passing the night with those RGs like the others, I spent a lot of time talking to her. She told me their story …"

Mike was about to ask him what he meant by these words, but he thought better of it. After all, that was none of his business. And then he was so tired that the only thing that really mattered to him in that moment was a good sleep. "I would like to sleep. Do you think it is wise to get asleep here, inside a replicator?"

"Ja, in the belly of the whale, like Jonah", Campos answered. "Yes, I will switch it off, and program the shuttle's computer to wake us if it detects any activity from our guest".

Mike stretched out on his seat to get some sleep. He was so tired he fell asleep immediately.

Campos checked again to ensure that the replicator was off and that no activity was going on, and then also prepared to go to sleep.

25

Mike slept for nine hours. When he eventually woke up he saw that Campos was already working.

As soon as he realized that Mike was awake, Campos told him "I have tried to switch on all the replicator's systems, and the machine answers correctly to all my commands. Try it yourself".

"Give me time to eat something first", Mike answered, getting up and taking some emergency rations from a locker. But he was so eager to try to control the replicator that after a few minutes he was back at his station. With Campos' help, he switched on the engines and set course toward a small asteroid that was marked on the shuttle's maps at a distance of a few million kilometers from their position. When the replicator started to speed up, Mike was overawed. They had worked so hard in the last few days, hoping to reach that result, but only now did he realize what their achievement meant.

He was at the controls of the most powerful spacecraft a human being had ever flown, of the most powerful war machines humans had ever conceived. The aliens who had built it had never invented warp drive. They seemed to have put all their hopes in artificial intelligence and huge automatic spacecraft, instead of developing faster and faster machines, that would allow them to travel among the stars in times that were consistent with a human lifespan.

"Takashi, you were right when you mentioned those arguments three hundred years ago, when at the beginning of the space age the supporters of human exploration opposed those who supported robotic exploration. Eventually we understood that humans had to explore space in person. These aliens, on the other hand, put all their hopes on robots, on their Von Neumann machines", Mike said turning toward Campos.

"Now that doesn't matter any more. We can put a warp drive on this replicator and turn it into a starship. And with a pressurized hull and some artificial gravity, it will be perfect to carry humans", Campos replied.

And that starship was Mike's. All right, he had to share his reward with Campos, but the shuttle was his, and he had paid a fortune for it. The largest share of the reward was his. He was even richer than before. Immensely rich and the owner of the most powerful starship humans could conceive.

He tried to concentrate on what he was doing, and turned to Campos, who had started to explore the memory banks of the replicator's computer. "Let me see what you are finding in that bottomless pit", he said.

"Of course. I am connecting your screen to mine, so that you can see everything I find".

Slowly the contents of that huge database scrolled past on the screen. The memory banks held everything the replicators had discovered as they roamed throughout the galaxy for millennia.

The first thing they wanted to understand was why these machines behaved so ferociously to all organic life. Actually, they really behaved as if they hated organic life, if a machine could ever hate. To such an extent that their mission seemed to be aimed at locating it and destroying it. What was hard to understand was why machines built by organic lifeforms should behave like that.

Meanwhile, they had reached the asteroid, a small rock, a few tens of meters in size that travelled on a well known orbit. As soon as it was in range, Mike operated the plasma guns and vaporized it with a single shot.

When he was sure he could control all the machine's systems properly, including the weapons, Mike changed its trajectory and began an intercept course towards the *Morning Star*.

Once he was able to operate the replicator's sensors as well, he realized why those machines had caused so many tragedies. Even at that distance he was able to analyze the contents of the ship and to see, one by one, all the organic lifeforms that were on board. Now that their cloaking devices were off, he also saw the emergency capsules and could tell how many people were on board each of them. He thought he could identify the capsule that carried his friends. 'One of those dots is Ann', he thought.

He checked the intercept course with the ship.

"Takashi, we have twenty-eight hours before we get to the ship", he said. "Apart from eight hours for sleep and two hours for eating and other essential things, we have sixteen hours to study that machine's memory store. Do you think that is enough time for us to get a clear idea of its nature?" He was formulating a plan and wanted to know whether it was feasible.

Campos was already at work, but it took longer than he expected to obtain, from the oldest memories, an image of the beings who had built the first replicator. They were humanoids, quite different from the humans from Earth, but nevertheless also quite similar to them. Their biology was based on carbon compounds, they breathed a mixture of oxygen and inert gases and fed upon lower forms of life, both vegetable and animal.

They found documents dating back to when the replicators had been first built. From the few images of the sky, seen from their planet, Mike tried to understand when that had happened, but he realized that they needed a much deeper study of the material they had found to get an answer. Anyway he had the impression that it must have happened at least a million years ago, and perhaps even much earlier.

That civilization had successfully colonized its own planetary system, which was in a zone of the Milky Way closer to the periphery than the Sun's location, in the Perseus arm of the galactic spiral. The distance between the stars in that region was considerably greater than around the Solar System, so that interstellar travel was even more difficult. Their science had reached a stage not very different from that of Earth at the beginning of space travel, and they had something similar to relativity. They were obsessed by the impossibility of creating a true space-faring civilization, additionally because they were aware that their star was aging.

They had placed all their hope on robotics and had built huge machines able to reach the nearby planetary systems in a few hundred years, and to build copies of themselves using materials mined from asteroids. Their final goal was to use their replicators to carry their own frozen embryos to planets in nearby systems. After surviving the very long journey, the embryos would develop in purpose-built artificial wombs to colonize the nearby planets.

Mike realized that the same ideas had been put forward on Earth about three centuries earlier, but then the development of warp drive had made them obsolete.

Anyway, they had not succeeded in colonizing the neighboring planetary systems like that. Soon they realized that random errors would be present in the replication process and the new generations of the robots would not be identical to the original ones.

When they realized that some of the replicators that were produced in distant systems might become dangerous, they made the fatal mistake of building some armed replicators with the goal of locating and destroying the machines that had the most dangerous mutations. But these also started mutating and, what was worse, the changes in the software started diffusing as well, through the transmissions the replicators exchanged with each other. The final disaster occurred when an armed replicator mutated in such a way that, rather than attacking other replicators, it attacked organic forms of life.

The machines eventually entered systems that were inhabited by other intelligent forms of life, which they attacked. The aliens reactions acted as a sort of selection mechanism. Because the harmless replicators were easier to destroy than the dangerous ones, eventually only the latter survived. Over the millennia, only the replicators that destroyed organic life had spread out, causing the extinction of life, at least of complex life, in the zone of the galaxy the they had occupied.

Apparently, the first intelligent species that became extinct was the species that had created the monsters.

Mike thought of the command sphere of the replicator they had captured. Even there, the thousand small mutations that occurred over the millennia were clear, and testified to the fact that no longer did any living being have any control over one of those machines. Their creators thought they could keep direct control of the giant spacecraft that they had built, but were able to do so only for a few generations.

Mike and Campos were still deep in the study of that ancient alien world, when a voice brought them back to the present: "This is Captain Romero. Mike Edwards, are you still in control of that replicator?".

Mike was ready to face the present. He had still his mind full of the enormous catastrophe caused by a reckless civilization that had tried to secure its future, using inadequate technological means, causing not only its own destruction, but also creating havoc for a still undefined number of other civilizations.

26

"Mike Edwards speaking. We are definitely in full control. We plan to rendezvous with you in less than an hour", he answered, after checking his instruments.

The ship was still far off, but was clearly visible on the screen. The station's emergency capsules had already reached the ship and all the survivors were on board. Mike was sure that the life support system, although damaged, would still be able to ensure the survival of everyone for a long enough period for them to reach a safe place. He continued his approach, thinking of what those on board must feel, seeing a replicator getting so close to the ship. He had said that it was now harmless, but he was sure they still had many doubts.

When he was close to the Morning Star, he saw a shuttle leaving the ship and coming straight in his direction. Joe appeared on the communication screen. "Mike, Ann and myself request permission to come on board".

"Permission granted, don't worry. Get inside the main structure of the replicator and dock to my shuttle. You can connect the two airlocks, so that you can come on board without having to use space suits. Be just careful of the maintenance robots: there is a lot of activity on board fixing the damage we caused with that warhead".

He followed the maneuver on his screen. But he was not really worried. There were many maintenance robots around and although flying through the structure of the replicator was no easy task, Joe, before getting his job on New Shanghai as flight controller, had worked as the pilot of a tugboat in a shipyard. He knew how to fly in cramped spaces. 'After all the times we had to listen to his adventures recovering damaged ships with his tugboat, now it is time for him to show us what he is actually able to do', he thought remembering the long evenings spent at Steve's talking and drinking beer. Now that past seemed to belong to a previous life.

After a while he heard the noise of a shuttle docking and the screen came on again. "Ann wants to speak with you alone", Joe said. "See you later". Mike got up and went to the access door. He was nervous: he didn't know how Ann had reacted to his leaving the station and above all to the fact that he had taken Lulu. Now he was telling himself that he did it to make her jealous, but he knew only too well that it wasn't true.

The door opened suddenly and Ann rushed in. She hugged him and remained there without saying anything. Mike realized she was crying and clasped her in his arms.

They remained there, without saying a word. Then Mike gently pushed her away to look in her face and froze, staring at her. Even with her eyes swollen with tears she was more beautiful than he remembered.

He turned and saw Campos who had entered the airlock and was going toward the outer door. "You have many things to talk about … I'm going to the other shuttle, so you can speak without people around", he said.

"Thanks Takashi", Mike answered. Then, turning toward Ann, he went on "Come … I think I owe you some explanation". He led her to the cockpit and sat down on the pilot seat, pointing out the co-pilot seat to her. But instead, Ann sat on his knees, putting her arms around his neck and her head on his shoulder.

"Mike, Romero asked us to empty your cabin and we found Lulu's body". Ann spoke without looking at him. "They told us everything, how you used a robot to disable that replicator. I am sorry you had to kill her …"

"Ann, that wasn't Lulu's body. That was the outer casing of a machine. A machine I used to save us all. I am not the slightest bit sorry for what I did".

"But Lulu was beautiful, even as she was without her insides, she was much better than me. And to see her in like that … I was sorry for her …"

"Ann, for one thing she wasn't more beautiful than you, not by any means", he interrupted her. "And then she was just an object, little more than a statue.

She was a useful object, useful as a weapon, I mean. She saved us all. Now let's forget about her. We cannot be grieving for a robot".

"But you loved her", she said, starting to cry again.

"Well, at one time I hoped to make something more than an object out of her. Perhaps I let the theories of that heretic Father Dubois influence me. All those stupid idea about complex robots having a soul. No, Lulu had no soul, not even intelligence. And this bloody replicator is just a machine too. At least RGs are designed by someone for a task, perhaps a disreputable task, but a precise task nevertheless. Replicators on the contrary are just the end product of uncontrollable evolution. They are just predatory animals".

Saying this he realized that, at the moment, Ann was not the slightest bit interested in a conversation about replicators, so he went back to the original subject. "Ann, Lulu is gone and I have realized that RGs are only objects, and dangerous ones at that, owing to their human appearance and behavior. Or rather, RG 46/G is gone, not Lulu. Lulu never existed. Please, forgive me", he concluded, looking her straight in her eyes.

"I am the one who needs to be forgiven", she said, "I didn't realize what the hunt meant to you. That is your world, your destiny. You are a hunter, a great hunter, the greatest of them all. It is enough to see how you have saved us all from this replicator. I am just a silly girl that allowed fear to rule her. I had no right to ask you to give up all this."

They were interrupted by Romero's voice from the screen. "Captain Edwards, we are requesting a tow to the closest spaceport".

Amazed by his respectful tone, Mike turned toward the screen. "Sorry, Ann", he told her in a low voice, lifting her from his knees and placing her carefully on the co-pilot's seat, "I cannot fly this crate with you sitting on my knees".

He switched on the transmitter. "Certainly, Captain Romero, we can tow you to the second planet. It will be expensive, but we can do it".

Romero was perplexed. He made a face that Mike found quite amusing. "And what would be the fee for towing us to the second planet?", Romero asked.

"I could do it for half a billion, what do you think?" Mike answered, unable to suppress a smile.

"What? Half a billion? I could get a tow to Earth for half that money. And then there are rules and laws regarding rescue operations". Romero was getting furious.

"As far as I know, such rules apply to rescue operations between human starships, they don't involve alien replicators. I'm sorry, but I don't believe that this replicator could settle for a lower price. If you prefer asking for assistance from another ship, you can try to call them. It's a pity that you will have to stay here for a few months, waiting", Mike answered. Then he added: "after all half a billion is not that much: it's the cost of a shuttle, and one that is not in very good condition too …"

At that point Romero understood. "Give us that bloody cable, you son of a bitch. I'll cancel your debt and we will be even", he answered.

Mike turned to Ann and realized that she was laughing. He took a handkerchief from his pocket and gave it to her, and she started to clean the tears from her face.

Then he called Joe: "Can you please get a tow cable, attach it to the back of the replicator and take it to the ship?"

"Yes, I imagine you want to be alone for a bit longer. I've done this for years, I can get a ship ready for towing again", Joe answered, starting to disconnect the two shuttles.

"When everything is OK, you, Steve and Takashi should come on board. I have work for you all, an offer I believe you will not turn down".

"Well, boss. If I understand you correctly, from tomorrow we will all be working for Edwards Enterprises".

"Yes, but what you cannot imagine is the purpose of the company …", Mike finished, turning toward Ann.

"Now that the work is done, why don't you come back here?" he asked her.

She got up and came back onto his knees. "He got what he deserved, that son of a bitch. From the moment I got on board, he didn't stop staring at me. You properly gulled him …"

"That's the second time in two days. I am sorry that the other time I had to cheat the Hunter as well … By the way, how is he?"

"In a bad condition, but he will survive. They say he will retire. From now on you will be the most famous hunter in the Galaxy. But what did you mean by saying that you lied to the Hunter?"

Ann was looking at him in amazement, waiting for an explanation.

"They told you that I left on a shuttle to fight the replicator because with the propulsion down, there was no other way to save the ship", he started.

"Yes, of course, it was a desperate attempt, but it was the only thing you could do."

"This was what I told the Hunter, and he was in such a poor state that he believed me. But it was a lie, a lie as big as a replicator".

He could see her amazement. "Yes, think about it. Our propulsion was down, but we still had four shuttles. If we had launched all the shuttles, they could have towed the ship out of danger, and the replicator, once it had destroyed your station, would not have been able to find us any more. We could get away, perhaps slowly, but safely".

"So you took that risk and sacrificed Lulu to save us!", Ann exclaimed.

"Yes, of course, to save you, and also the others. But you didn't get it?" Ann hugged him again. "Madame had told me that you bought Lulu just to play on my jealousy to make me accept you becoming a hunter. But I didn't want to believe her …"

"I am sorry, but to be honest, things were not like that", he answered earnestly. "I wish I had done it for that reason. But then I was so angry with you and I believed I could make something more than a robot out of Lulu." He was silent for a short while, then added "Can you forgive me?"

"Of course I will", she answered with a smile. "Also because all of us on the space station would have been dead by now, if you had not had that robot with you. As you can see, this is the second time that Lulu has saved my life. I suppose I am most ungrateful, but I am quite happy I got rid of her this way", she added with an expression of mock contrition.

"And then you must forgive me. But now you will not get rid of me that easily. Your destiny is to be a hunter. With all the money you have now we will buy … you will buy … the best hunting ship and you will be the most famous hunter of all time. And I will be with you. After all, now I am no longer so afraid of replicators", she added, realizing that at that very moment she was right inside one of them.

"No, Ann, you were right. Hunting replicators is useless. There is no sense in facing these huge war machines with a ship that does not have enough armament, without appropriate equipment, and do it just for money as hunters do. It's a dangerous and meaningless business.", Mike answered.

"Yes, but as you told me, there is nobody who is ready to build a true fleet to wage a full-scale war against them, with adequate means. The government of Earth is too short-sighted to do it since the danger to Earth is too far away. And the colonial governments are too weak and poor in this part of the Galaxy. And what is more, while replicators advance, this part of the colonized zone becomes less and less populated and poorer, and the governments become weaker", she said, hardly believing that she was the one saying those words.

"But there are not only the governments, Ann. Sure, they have done very little up to now, and they will never do much more …"

"But then, who? Who could build a powerful enough fleet …" she interrupted him.

"Me. Or rather, we … That is Edwards Enterprises. Do you realize that this replicator is a huge shipyard, able to turn out a replicator every sixty days? We will reprogram it, so that it will no longer be a Von Neumann machine, but an automatic shipyard directly controlled by us. It will produce powerful warships, armed with plasma jets, able to destroy replicators safely, and huge freighters, able to carry thousands of people to the systems we will open up to colonization. From this replicator's memory we can obtain detailed star maps, with information about millions of star systems. We will be able to receive the transmissions broadcast from the replicators scattered throughout the galaxy. We will mop up this part of the galaxy, we will free it from replicators, and we will give it life again with civilization from Earth".

After a pause, he went on: "I think you are right. My destiny is to be a hunter. But not to participate directly in the hunt. Within a few years we will have hundreds of people hunting on our ships, hundreds of colonies living thanks to the supplies carried by our freighters ..." He suddenly stopped, realizing he had been carried too far. Above all, the tone of his voice was out of place. He lowered his head and added quietly: "Something of that kind, anyway ..."

Ann hugged him fiercely. "That's great, Mike. At last people in these systems at the edge of the colonized zone will have a normal life ..."

The door suddenly slammed open. It was Campos, pale as he had never seen him, not even during the most dramatic moments of the attack.

"What's going on, Takashi? Didn't you go with Joe?", asked Mike, both worried and upset.

"No, I remained in the hold, studying the files I saved from the replicator. Take a look at this", he said, handling him a small screen.

At first Mike could not understand what he saw on the screen, then he realized that it was one of the transmissions that the replicators had exchanged. Despite the huge interstellar distances, the news that there was an intelligent species in that part of the galaxy had already spread. Within a few months, the avant-garde of the packs of these dreadful machines would, after travelling for years, reach the frontier of the colonized zone, from where they could attack the inhabited planets.

The silence that had filled the room was interrupted by Joe, who was clearly unaware of the news: "The cable is in place. You can start, Mike".

"Be careful, I'll put the cable under tension", Mike answered, as if he just woken from a nightmare, seemingly not reacting to what he had just seen. The replicator started moving slowly towards the second planet, while the tow cable grew taut, towing the Hunter's ship.

Part II

The Science Behind the Fiction

Humankind on the Verge of Becoming a Spacefaring Civilization

Introduction

The action is set in the year 2328, in the system of the double star BD–05 1844 (or Gliese 250) at 28.4 light years (9.2 parsecs) from the Sun[1]. The primary star, BD-05 1844 A is an orange-red dwarf star (K3V), with a mass about 80% of the mass of the Sun but a luminosity of only 14.6%. BD-05 1844 B is a red dwarf, (M2.5V), with 50% the mass of the Sun and only 0.58% of its luminosity. Their separation is about 500 Astronomical Units.

Its apparent magnitude is +6.58 and thus it is essentially invisible to naked eye observers. No exoplanets have yet been discovered orbiting this double star, so the planets (and their moons) mentioned in the novel are fictitious.

The model assumed to assess the science and technology at that time in the future is an evolutionary model: Technological advances occur through a slow refinement of the technologies and of the scientific theories that underlie them. Scientific revolutions are rare and technological revolutions even rarer. Sometimes a number of fields of technology may have a rapid development, followed by a period of stasis in which little happens, while other fields begin a dramatic advance. An example of this is the period between 1935 and 1965 for aviation and then space travel. In 30 years humans passed from propeller-driven biplanes to supersonic jets to and rockets that allowed them to land on the Moon. The nuclear rockets that would have allowed humankind to become a spacefaring civilization were even tested on the ground. Then in the following 45 years little happened, or worse, there was a setback. As we know, a heavy lift rocket such as the Saturn 5 no longer exists, the Space Shuttle has been scrapped, the supersonic airliner is no longer operational, and nuclear rockets have not materialized.

When in 2004 a return to the Moon was seriously being considered by NASA, the Constellation program, which included the launchers Ares I and

---

[1] It lies in the east central part (6:52:18.1-5:10:25.4 for Star A and 6:52:18-5:11.4 for Star B, ICRS 2000.0) of the constellation *Monoceros*.

V and the spacecraft Orion, was initiated. It was mainly based on technologies similar to those that 40 years ago allowed humans to reach the Moon but, mainly for cost reasons, the whole program was canceled in 2010.

While aerospace technology has not made the quick steps forward that were predicted, we have had striking advances in computers, electronics, cell phones, etc. True innovation is usually unpredictable. For instance nobody predicted the diffusion of ICT such as personal computers, cell phones and the internet, while many advances in other fields, which were predicted to happen in the near future, never materialized.

The science and technology described in the novel are not very advanced, except for assuming a single 'scientific revolution' that fuels a 'technological revolution': the warp drive. This propulsion device allows humankind to start a true interstellar spacefaring civilization, expanding in a sphere with a radius of about 9 parsecs centered on the Sun by the time in which the novel is set.

The main scientific aspects that enter this novel are related to robotics: some characters actually are robots, and the villains of the story are robots, those still hypothetical robots usually referred to as *Von Neumann machines*.

Space Travel

The basic assumption I make is that little useful work can be done in space using chemical propulsion. After the race to the moon of the 1960s and 1970s, the present stagnant situation is assumed to have lasted until about 2020, when space exploration resumes thanks to private investors developing space tourism and, later, asteroid mining. A further assumption is that nuclear propulsion, derived initially from the old 20th century studies [1, 2] and above all from the nuclear rocket built and tested on the ground as a part of the NERVA program, allows humankind to reach Mars and nearby asteroids in a reasonable time, and is also instrumental in making faster and cheaper journeys to the Moon.

Mars is assumed to be a barren desert, with no life at all, and plans for terraforming the planet are drafted as soon as the absence of indigenous life is ascertained. In the novel, the expansion in the Solar System proceeds with slow improvements until the end of the 21st century, with the noteworthy application of nuclear fusion to space propulsion. This results in opening up the main asteroid belt to exploitation.

Theoretical ideas on propellantless propulsion and warp drive were advanced at the end of the 20th century [3–5]. In particular, the Breakthrough Propulsion Physics (BPP) program [6], which focused more on physical and mathematical aspects of advanced space propulsion than on applications, was active between 1996 and 2002. Its goal was to lay out the scientific foundations of what could become a new technology some decades from now—*to perform credible progress toward incredible possibilities*, as the catch phrase of the program said. In 2013 NASA resumed studies on a warp drive, and in the novel it is assumed that new advances first allowed a better theoretical understanding and then the development of a technology allowing FTL (Faster Than Light) interstellar travel.

Propellantless propulsion (or space drive) is assumed to be achieved first, allowing humankind to draw up plans for interstellar colonization journeys at speeds lower than that of light. However, before these plans could be implemented, I make the assumption that the first warp drive starship was launched and a FTL probe sent to Alpha Centauri. Colonization of nearby exoplanets could thus be started with journeys lasting months instead of many years or centuries.

A further assumption is that a technology for controlling gravity may be obtained from the same hypothetical development of physics that enabled propellantless propulsion. Artificial gravity could thus be created on board starships (on space stations the same goal is obtained by rotating the station to save energy), with the added advantage of compensating for the high accelerations needed to reach speeds comparable with the speed of light in a reasonable time.

Such fast starships (in this fictional world as well as in the real world) would have to be provided with shields to prevent damage from collisions with micrometeoroids and other objects. The same shields can also be used as a protection against weapons, something needed—in the novel—since the rapid expansion of human civilization at rapidly increasing distances from Earth produces an unstable situation in the frontier zones where the novel is set. There, encounters with unfriendly people—humans, since no aliens will be encountered at those (astronomically close) distances from Earth—are by no means rare. At the time at which the events described occur, the presence of hostile replicators (see below) will disrupt the peaceful order of society, increasing the probability of such encounters.

Warp drive requires huge quantities of energy and in the novel a solution is found for storing energy on board. Antimatter is produced from deuterium and helium-3 mined from the atmosphere of gas giants and burned in fusion reactors. Antimatter is then stored in huge space stations orbiting the same planets and used as an energy medium to power starships.

Using warp drive, journeys among systems separated by a few parsecs may take some weeks. This duplicates the situation on Earth in the 17th and 18th centuries, when transoceanic travel was slow and costly—but possible, nevertheless, thus allowing empires spanning different continents to be built.

Colonies can therefore be built on some extrasolar planets but, since in the novel extraterrestrial life is assumed to be rare, open-air settlements will be started on just a few planets because of the lack of suitable biospheres supporting an atmosphere rich in oxygen. Most of the planets need thus to be terraformed, and this process is assumed to be started in several places, mostly by private terraforming and space engineering companies. One of these companies, partially owned by the Chinese government, is said to start terraforming operations on a terrestrial planet orbiting Gliese 250 A in 2270.

At present, the science and technology that will allow terraforming planets are still in their infancy and it is uncertain how it may proceed. What is certain is that the complexity and the approach required are strictly dependant on the characteristics of the planet. In the case of Mars, for instance, terraforming operations can be divided into two phases—increasing the atmospheric pressure, perhaps by heating the surface, and making the air breathable [5, 7]. The whole operation may take a very long time, but some estimates as short as 500 years have been proposed [7]. Shorter terraforming times may be made possible by the use of nanotechnologies.

Recently, the idea that even small and airless worlds like the Moon may be terraformed has been considered. Owing to its small mass an atmosphere cannot be made stable, but the time needed for the Moon to lose an artificial atmosphere may be so long that a modest amount of gas released continuously would be able to compensate for the losses.

In the novel, the terraforming processes on a number of explanets, like the one assumed to exist in the Gliese 250 system, are described as being under way. However, since it seems that the number of exoplanets is very large, some selected planets with favorable characteristics that can be terraformed in a short time are assumed to exist.

Another assumption is that progress in materials will allow the construction of space elevators [8] within the timeframe described in the novel. However, owing to the cost of such infrastructure, only the Earth and a few colonized planets are assumed to have traffic between the surface and space, allowing a space elevator to be cost-effective. Moreover, the cost of space transportation is assumed to be quite low, mainly thanks to the use of nuclear propulsion, not only beyond Earth orbit, but also in the last part of the satellization run. Under such circumstances it may be expected that the volume of traffic that can justify a space elevator is very high indeed.

Another idea that permeates the novel is that the starting of a spacefaring civilization will have a strong effect on humankind. Interbreeding will cause human races and differentiated cultures to almost disappear and a single human type will start to emerge. This is reflected in the names of the characters, which are often a mixture of what today are names and surnames from different cultures and nations. However, new differences are assumed to start emerging between people living on high- and low-gravity planets, or in high- or low-pressure atmospheres [9]. This latter process is much slower than the effects of interbreeding, and at the time the novel is set it is assumed to be still marginal.

Astrobiology

The basic astrobiological theory followed in the novel is what is usually called the "Rare Earth Hypothesis": Life is fairly common in the universe, but only at the level of its most elementary types [10]. The possibility that no complex life exists within 10 parsecs from Earth is not surprising, as even the most enthusiastic supporters of SETI (Search for Extraterrestrial Intelligence) would readily agree. By introducing very optimistic numbers into the Drake equation, the result is an average distance between intelligent species of about 200 parsecs. The astrobiological background of the novel is thus consistent with today's mainstream astrobiological thinking [11].

In the novel, then, a number of planets in the range of 9 parsecs from Earth (i.e. within the colonized zone) are assumed to possess an elementary kind of life which, in some cases, has transformed the atmosphere, enriching it with oxygen and making it breathable (or almost breathable) by humans. This is the case, for instance, of Ceres, a fictitious Moon of a planet orbiting Gliese 250 B. This moon could be colonized easily, without the need for being terraformed (unlike completely sterile planets, which are assumed to be the majority). As a consequence, since terraforming is a long and costly process, particularly in the case of hostile planets, the colonized zone is said to contain few inhabited planets. Most settlements are described as small space stations or mining bases on asteroids.

As already stated, nothing is known about possible planets in the Gliese 250 system. In the novel the system is assumed to contain several planets, two of which are particularly interesting, and asteroids. The first of the interesting bodies is a large terrestrial planet, orbiting the A component, which hosts no life forms and is being terraformed. This planet lies in the habitable zone of the star, quite close to it, but is not gravitationally locked so that, once terraformed, it may become an important center in the frontier zone.

The second is the satellite of a giant planet orbiting very close to the B component of the star. It has a very primitive biosphere, consisting of just bacteria, that have enriched its atmosphere with oxygen and made it directly habitable. At the time in which the action of the novel is set, the body has been settled and terrestrial life forms introduced. The ethical problems linked with the subsequent, almost certain, extinction of the indigenous forms of life are not discussed, but it is clear that the company that owns the body, and wants to use it as a logistics center for terraforming the other planet, does not care much for the subject. It set up a sealed enclave in which the local forms of life are preserved for scientific reasons, but that is all. In the frontier zone, at 9 parsecs from Earth, there is nobody who can enforce rules for the protection of bacterial forms of life. And, after all, we know that humankind has caused the extinction of bacteria or viruses, such as the virus responsible for smallpox, without too many ethical qualms!

No planet in the colonized zone (nor seemingly in the known Universe) is believed to contain higher forms of life. For centuries the common opinion was that the only place in which complex life, and then intelligent life, could evolve, was Earth. This belief is shaken in year 2294, about 30 years before the time at which the action in the novel takes place, when mysterious self-replicating robots are found: if they are artificial, as most people believe, then some alien species must have built them. However, some characters in the novel still oppose this theory, suspecting that the *replicators*, as these robots are called (see below), evolved by themselves, without being built by anyone.

Robotics

Robots and artificial intelligence

The term 'artificial intelligence' was first used in a conference held in the summer of 1956 at the Dartmouth College, New Hampshire, and has been widely used ever since. Although no commonly agreed definition of artificial intelligence (or, for that matter, human intelligence) exists, there is general agreement in accepting that it consists of a machine imitating human intelligent behavior. The well known *Turing test* comes closest to a definition: a machine is intelligent if it is impossible to distinguish it from a human being during an interaction based on exchanging messages on any subject.

Two methods have been employed to reach the goal of creating artificial intelligence. The first consists in writing dedicated software running on conventional, although very powerful, computers, able to manipulate symbols following well-established rules. The basis for these attempts is the assumption that intelligence is based on algorithms that perform logical operations by manipulating symbols. The human brain is thus considered to be a biological computer, and the human mind is the result of some sort of software running on it. This method supposes that intelligence and even consciousness can be obtained by using a non-biological computer, provided it is powerful enough to run suitable software.

The second approach is the neural one, based on the construction of a network of artificial neurons, simulating the structure of animal, and also human, brains. The Artificial Neural Networks (ANN) so devised operate not by running programs but by learning.

These two approaches are not as different as they look, because the operation of neural networks may be simulated on a computer, i.e., it can be reduced to software running on a conventional machine. In this way the second approach seems to reduce to a particular case of the first.

At any rate, the neural approach seemed to be haunted by unsolvable problems and at the end of the 1960s it seemed to be a dead end. The algorithmic approach, on the other hand, seemed to obtain encouraging results.

By the mid 1980s the neural approach regained momentum when the above-mentioned problems were solved, but the goal of building intelligent and conscious machines proved to be much harder than predicted. In the 1960s it was a common opinion that by the year 2000 there would be an established technological result. At the time of writing, in 2013, artificial intelligence seems still to be far away and many scientists cast doubt on its feasibility, at least with present-day technology [12].

A tentative scheme of the path leading from inert matter to intelligent and conscious systems is shown in Fig. 1 [13]. According to this scheme, a material system, able to manipulate energy, is a dynamic system. If it can also receive signals and manipulate information, it may be considered as an automatic device, and so on. It is debatable whether the decision box should be over or under the knowledge box. Here it is assumed that a being (living or robotic) can take decisions reacting to the inputs from the outer world even without building an internal model of it. This point has been controversial, but here it is assumed that a positive answer to this problem is realistic. Moreover, sometimes it seems that there is no complete agreement on the meaning of the terms in the boxes, so that the answer depends on the exact interpretations of what the words 'knowledge' and 'decision' mean.

The boxes on the right should not be considered as separate steps in a ladder, but rather as levels in a continuous evolutionary process, and there is an infinite number of shades between each of them.

Telemanipulators, i.e. remote controlled agents able to perform well-determined tasks, are without any doubt at the second level—they are *automatic systems*, as are many other automatic machines of various kinds. In many cases, telemanipulators display some form of limited autonomy, being able to take low-level decisions, while being controlled by humans for higher-level tasks [13].

The earliest manipulators, used for the preparation of radioactive materials, were purely mechanical devices, consisting of an arm and a gripper, able to

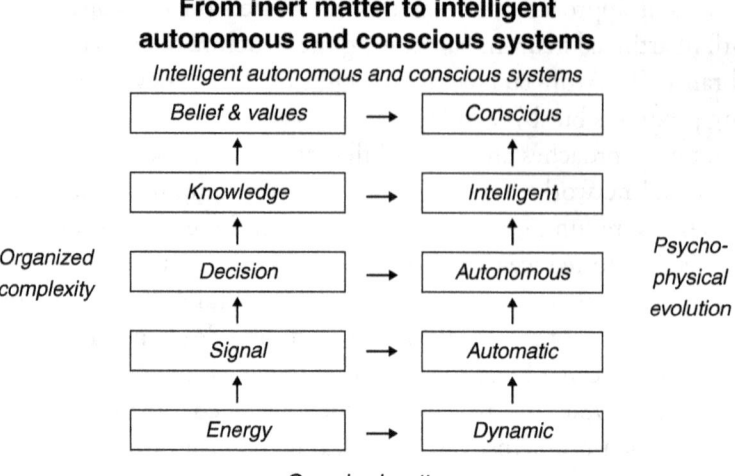

Fig. 1 Tentative scheme of the path leading from inert matter to intelligent autonomous and conscious systems. [13]

duplicate exactly the motion of the arm and hand of a human operator. Later, an increasing number of movements were performed under autonomous control, with the human controller simply taking decisions at a higher level. An analogy is that of the gearbox in a motor car. In basic transmissions the driver has to control the clutch and the gear lever, supplying the power needed to perform the action. In semi-automatic transmissions the driver takes the decision about which gear to engage and the device operates the clutch and the gearbox, using a source of power. In fully automatic gearboxes it is still the driver who controls the speed through the accelerator pedal, and in so doing causes changes of gear. Here all low-level decisions are taken by the device, but the high-level decision is still taken by the driver in real time.

To qualify as a true robot, the device must possess a good degree of autonomy, being able to perform its tasks without direct, real-time, human intervention. A robot must also interact with the environment and perform its tasks in a flexible, easily reprogrammable, way. Thus it belongs to at least the third level in Fig. 1.

Ideally a robot should be even more autonomous from human intervention and should also perform in an intelligent way or even be conscious. Both these characteristics are still far from being present in actual robots.

Speaking about space exploration, an important point is how much autonomy, and thus artificial intelligence, is needed for this task. Generally speaking, it can be stated that the autonomy of robots must increase with the distance from the human controller at which they operate. While it is possible to conceive telemanipulators for all tasks to be performed on the Moon, the autonomy required for Mars exploration (at least until humans are present on that planet or on its satellites) must be of a higher degree. The distance of satellites and planets in the outer Solar System is such that unmanned exploration requires true robots. The distances at which unmanned devices will operate, when interstellar exploration is undertaken, may make it necessary to resort to intelligent machines, in the sense defined above.

Strong AI is based on the assumption that all human characteristics can be duplicated by machines and consequently that they will not only be intelligent, but they will also possess a true mind, with related consciousness. This is, however, an unproved statement and, particularly the last part, quite a controversial one.

A basic assumption in the novel is that the strong AI hypothesis is not supported, not only by the lack of serious evidence, but not even by hints of it. Thus, since thought (and even more, consciousness) is not an algorithmic process, present technology and the technology foreseeable for a medium-term future will not permit the construction of thinking, self-aware machines. None of the robots of the novel could pass the Turing Test, and they are basically machines with an evolved control system but nothing more.

As consequence of the limitations of AI, severe difficulties are encountered in robotic space exploration. As stated above, interplanetary, and above all, interstellar exploration needs highly autonomous robots, able to take decisions without the help of humans. Advances in space travel make human exploration missions increasingly expedient, and the combination of low-cost access to space and efficient space propulsion strongly limits the role of robots in space exploration, except for their use as 'astronaut assistants' or teleoperators in dangerous or difficult places.

Humanoid Robots

Robots are almost a commonplace in science fiction, and the word *robot* was first used by the Czech writer Karel Čapekwho, in 1920, published his science fiction play *R.U.R. (Rossum's Universal Robots)*, dealing with artificial men built for performing work in place of human beings. He invented the word *Robota*, from the base robot-, as in *robota*, compulsory labor, or *robotnik*, peasant owing such labor. (Words that originated in science fiction, such *robot, terraforming* and many others, are now commonly used in the scientific and technological jargon). The robots described in the play are humanoid.

Originally a robot was a sort of 'artificial human', a mechanical slave that had a humanoid body, and sometimes a humanoid behaviour, not to mention the cases in which it also had a humanoid mind. For instance, the definition of *robot* from the *Random House Webster's Dictionary* is "*a machine that resembles a human and does mechanical, routine tasks on command* or *any machine or mechanical device that operates automatically with humanlike skill.*" However, other definitions are wider, and more realistic: the ISO (International Standards Organization) 8373 standard defines a robot as "*an automatically controlled, reprogrammable, multipurpose, manipulator, programmable in three or more axes, which may be either fixed in place or mobile for use in industrial automation applications.*" The ISO definition thus includes industrial robots, which are neither humanoid nor biomorph (nor bio-inspired). However, the robot remains, in the collective imagination, a sort of artificial human, and is the very essence of a bio-inspired machine.

For some applications, mostly when they are designed to work with humans, it may be expedient that the robot actually has a humanoid shape. For instance, a robot that has to move in a structured environment designed for humans may perform better if its overall shape is that of a human being. Other cases are robots built for studying animal or human locomotion or gestural communication: an example of this is Kismet, a research robot built at MIT by Dr. Cynthia Breazeal. It simulates emotion through various facial expressions, vocalizations and movement, and its aim is the study of non verbal communication [14] or those built for the *edutainment* market. Robots working in hospitals, attending to elderly people or children, may be more reassuring if they have a human look. Above all, they may convey messages through their facial expressions and body postures that are immediately and subconsciously understood by people—even people some of whose faculties are impaired, so that the interaction with humans is improved by simulating a humanoid shape and behaviour.

By using a humanoid (or at least animal) means of locomotion (legs instead of wheels), climbing stairs, getting on elevators, going through doors and in general moving in an environment designed for people can be simpler and a humanoid robot can use tools designed for humans: drive vehicles, etc. This is typical of personal robots designed for performing domestic duties: instead of owning a number of household appliances it may be more expedient to own a single service robot (Fig. 2) that can use the same hand tools that were previously directly used by humans.

Apart from the above-mentioned cases, the same functions performed by a humanoid robot may be better performed by a machine whose configuration is directly dictated by its tasks and not by the way nature solved the same problem. A 4-wheeled centauroid robot may be far more suitable to collaborate with humans under many circumstances, both within an unstructured and a structured environment, than its biped counterpart.

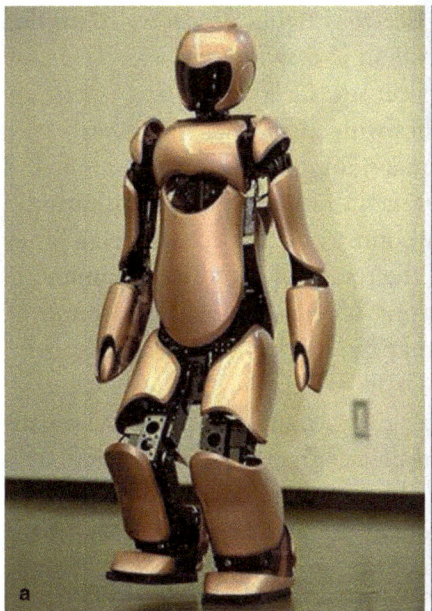 

Fig. 2 Domestic robots **a** e-NUVO, a small domestic robot, developed by the Nippon Institute of Technology, with Harada Vehicle Design, ZMP and ZNUG design, (From http://www.plasticpals.com/?p = 18810 with kind permission of Prof. Kensuke Takita, NIT, Tokyo). **b** Valerie, anthropomorphic domestic robot presently under development by Android World. This is perhaps closer to the RGs of the novel, although the manufacturers say explicitly it cannot be used in that way. (From http://www.androidworld.com/prod19.htm with kind permission by Chris Willis)

Recent studies (see, among other others [15]) assessed that the market for personal and domestic robots will boom in the first half of the 21st century and some even state that the personal robot industry will have the role that the automotive industry had in the past. However, the performance of the humanoid robots built up to now (the Sony SDR-4X, a small, child size, domestic and entertainment robot, with a vocabulary of 60,000 words, the Honda Asimo work robot, and some others) are still inadequate and they do not yet have a true market.

In the novel, the advances in the field of robotics are not assumed to be particularly fast. Menial tasks are performed by robots that are not necessarily fully humanoid, but have a shape that is adequate for their tasks. These robots have been around for decades, and are certainly not at the cutting edge of technology. At any rate, the place where the novel is set is a space station in a distant, frontier system, so if these robots were costly new technological gadgets, they would not be there.

The only true humanoid robots in the novel are the 'RG', an achronym for robogirls, and their male counterparts. As a character defines them, an RG is "seventy kilos of machinery, moved by electrohydraulic actuators controlled by a computer and covered with silicon flesh and synthetic skin". Their use is the most private type of personal service: sex. On the frontier, where few miners and adventurers take their families with them, their use is, in a way, encouraged to reduce the number of humans doing this job. They mimic human behaviour, at least as far as their programmers succeed in replicating it and, at least from some distance, may be confused with a human being. They too are consolidated technology and are the results of a long development.

At present, many predict that something of this kind will be around by 2020, and there are even predictions that by 2050 they will change completely the market in prostitution [16].

As already stated, most of the technological advances in the novel are assumed to have taken place on a slower timescale than is now predicted. This conservative attitude applies also to details: robots are essentially made with a mechanical body, although the advances in the field of materials allow the construction of stronger and lighter structures, powered by electrohydraulic actuators and controlled by a computer, smaller and much more powerful than present-day ones, even if much less than predictable from Moore's law, but with no revolutionary changes.

These robots in the novel are not intelligent, in the sense that they follow programs, albeit very complex ones, and they are definitely not conscious. At any rate, because they are built to interact with humans and to imitate human behaviour, they are much more complex than other robots, and their computers have the most advanced processing units and memories.

However, the idea that strong AI is viable is still shared by some of the characters who think that, when the complexity of a computer goes beyond a certain threshold, it starts to develop intelligent behaviour. The discussion about intelligent robots is thus brought to a higher, more philosophical (or, rather, theological) level, echoing the 'consciousness-complexity' law from Teilhard de Chardin's theology [17]. A character in the novel states that when a robot is complex enough to reach intelligence and then consciousness, it is endowed with an immortal soul.

Von Neuman Machines

Von Neumann machines are self-replicating robots, although Von Neumann called them Universal Constructors. In the novel they are called replicators: a term introduced by Drexler[2] [18] for artificial self-replicating systems based on conventional large-scale technology and automation. The idea is however

---

[2] Actually Drexler used the term *clanking replicators*, to distinguish them from self-replicating machines at the micro- or nano-scale, which he called *replicators*.

much older, since non-biological self-replicating systems were imagined in Samuel Butler's article *Darwin Among the Machines* published only a few years after *The Origin of Species* [19].

Since 1980, NASA forwarded the idea of using self-replicating factories to develop lunar and asteroid resources, but their use in space exploration is particularly expedient if interstellar exploration is performed by using slow interstellar spacecraft[3], unable to reach a speed that is a substantial fraction of the speed of light.

Interstellar probes must have an operational autonomy that is close to true intelligence. Provided that replicators endowed with sufficient intelligence can be built and miniaturized enough, they could take command of interstellar probes, guiding them to their targets. At that point all distinctions between the intelligent machine and the probe would fade, and it would be more appropriate to talk of a *Von Neumann probe*.

Once it reaches its target, such a probe could choose a suitable asteroid, a planet with a solid surface or a satellite, land on it and start building a copy of itself. A strategy for space exploration based on them has been proposed by Frank Tipler [20].

A Von Neumann probe could be launched towards a nearby star with a comparatively simple propulsion system. After several hundred years, or even many thousand years, it would reach its destination. The probe would land and start producing other probes, which would then leave that extrasolar system, heading off toward other nearby stars. Once its primary task of continuing the expansion to other star systems had been fulfilled, the probes would begin their scientific tasks, sending reports back to Earth. Eventually, most of our galaxy would be settled by these probes. According to Tipler, a single intelligent species could even begin to explore the whole Universe using Von Neumann probes.

It has been computed that through this strategy, using replicator probes travelling at a speed not higher than 10 % of the speed of light, a galaxy the size of the Milky Way can be explored in as little as half a million years [21].

Such intelligent machines might not just explore, but also reproduce organic life. The question is: how small and lightweight can a Von Neumann probe be? Thanks to rapidly developing nanotechnologies, it might be possible to build a very compact and lightweight self-replicating machine, but in the novel the more common opinion that replicators must be gigantic machines, true self-sufficient factories, is accepted. A comprehensive review—provided with a wide bibliography—about replicators can be found in the book *Kinematic Self-Replicating Machines*, by Freitas and Merkle [22].

---

[3] The term *slow interstellar travel* usually indicates interstellar travel at a speed lower than 1 % of the speed of light ($v = 0.01$ c).

Even when a Von Neumann machine is built, could we be sure that, after many replications of itself, errors would not creep in? After all, this is one of the mechanisms by which evolution creates new living beings. Will a probe programmed on Earth always perform correctly in the new environments that it will find in other planetary systems? Checking, or even modifying, the programming of the probe by radio from the Earth is possible only for the first few replications. Then the distances in both space and time become so large that everything must be done by the on-board artificial intelligence systems. What might be the outcome of such machines, once they stop behaving exactly as their builders envisaged, owing to random modifications of their genetic code?

In this way small replication errors (mutations) could accumulate and a Darwinian evolution[4] of these machines would occur. Since they are huge and powerful, they could become the Doomsday machines (Berserkers) Stephen Webb mentioned in one of his 50 solutions to the Fermi Paradox [23].

Another, more important point has to be addressed. Assuming that such intelligent machines can be built, is it morally acceptable to do so? Should self-replicating machines fill the Universe? That question has caused fierce arguments. Carl Sagan believed the answer to be no, and he stated that [24]:

> …the prudent policy of any technical civilization must be, with very high reliability, to prevent the construction of interstellar von Neumann machines and to circumscribe severely their domestic use. If we accept Tipler's arguments, the entire universe is endangered by such an invention; controlling and destroying interstellar von Neumann machines is then something to which every civilization—especially the most advanced—would be likely to devote some attention.

Frank Tipler's counter-answer is equally strong. If humankind abdicates that role, it will miss all chances of colonizing, first, nearby stellar systems and then the Universe. Humankind will betray its cosmic duty, and condemn itself to extinction. To quote his words: [20]

> This is a position of fear and ignorance, a definition by exclusion: that which is unlike me is not worthy of existence. A "person" is defined by qualities of mind and soul, not by a particular bodily form.

By Frank Tipler's reasoning, the dissemination throughout the Universe of Von Neumann machines may be considered as another aspect of that evolutionary process which produced humankind and which may in future pro-

---

[4] The first time replicators were mentioned, by Butler in 1840, they were associated with evolution and were thought to be an application of Darwin's ideas to machines.

duce other intelligent species to take its place. The ultimate evolutionary task of humans would thus be to create intelligent machines, i.e., to move the evolutionary line from beings based on the biology of carbon to beings based on the chemistry of silicon.

Although Von Neumann machines are linked with artificial intelligence, and in their version by Tipler they are truly intelligent and conscious robots, a much less advanced artificial intelligence may be sufficient to build replicators. These less advanced replicators—the only ones that are possible within the assumptions considered in the novel—may be even more dangerous than fully conscious ones, which might, after all, have a system of beliefs and values (Fig. 1) preventing them from destroying all organic life.

In the novel, an alien civilization, in some distant part of the galaxy and lost in the mist of time, tried to explore its nearby systems using replicators, which in time transformed into the above-mentioned Doomsday machines, destroying their builders and any living being they encountered in their expansion in space.

Glossary and Achronyms

AI Artificial intelligence

ANN Artificial neural network

Astronomical Unit Distance unit, precisely equivalent to 149,597,870.7 km (92,955,807.3 mi). It roughly corresponds to the mean Earth–Sun distance.

Apparent Magnitude (of a celestial body) A measure of its brightness as seen by an observer on Earth, adjusted to the value it would have in the absence of any atmosphere. The brighter the object appears, the lower the numerical value of its magnitude. An average naked-eye observer under very good conditions can see a star of magnitude 6.5, while the extreme naked-eye visibility limit is between 7 and 8.

BPP Breakthrough propulsion physics, an advanced research program run by NASA from 1996 to 2002.

Centauroid Robot Robot having 4 legs (or wheels) and provided with a more or less humanoid torso with two manipulatory arms.

Drake Equation An equation, introduced by the radio-astronomer and SETI specialist Frank Drake, yielding the number of extraterrestrial civilizations that can enter in contact with us. The coefficients entered into the equation are highly hypothetical, so that the equation is useful for understanding which parameters govern the phenomenon, but is unable to supply a reliable numerical result.

Exoplanet Planet orbiting a star other than the Sun.

Gravitational (or Tidal) Locking Situation in which a satellite orbits so close to a planet (or a planet close to a star) that, owing to tidal effects, the rotational and the orbital motions become synchronized. As a consequence, the secondary body always has the same hemisphere facing the primary body. (The Moon is gravitationally locked to the Earth).

ICT Information and communications technology.

Light year Distance unit, equivalent to 9.4607×10^{12} km (almost 10 trillion km). It is the distance light travels in one year.

FTL Faster than light. FTL travel is considered to be a violation of physical laws because, as a consequence of Relativity, neither matter nor information can move at a speed higher than the speed of light. However, a deeper understanding of physics seems to suggest some possible mechanisms that might be valid in this respect (*see* warp drive).

Moore's law A law, originally stating that the number of transistors on integrated circuits doubles approximately every two years. It has been formulated in different ways, and is now generally interpreted as a doubling of computer performance every two years.

NERVA Nuclear engine for rocket vehicle application, a highly successful program conducted by the U.S. Atomic energy commission and NASA, which demonstrated the feasibility of thermal nuclear propulsion (NTP). At the end of 1968 the latest NERVA engine, the NRX/XE, was tested on the ground with a total run time of 115 min and met the requirements for a manned Mars mission. The program was canceled in 1972.

Parsec Distance unit, equivalent to 30.857×10^{12} km or 3.26156 light years. It is the distance at which the Sun-Earth distance subtends an angle of 1 arcsec.

Propellantless Propulsion Also called space drive, is a hypothetical way of propelling a spacecraft without ejecting material as takes place with rocket engines.

SETI Search for extra terrestrial intelligence.

Space Elevator A structure attached to the Earth's surface and extending into space allowing people and goods to be carried beyond geostationary orbit without the use of rockets or any other form of atmospheric vehicle. It may be imagined as a cable, extended downwards toward the Earth from a geostationary satellite until it is anchored to the surface. Another cable is extended outwards to balance the weight of the former and another satellite (the outer station) is located at its end. Any object released from the outer station is thus launched towards the outer space. The total length of the cable is about 60,000 to 100,000 km. The cable is stressed well beyond the possibilities of any existing material, but it is believed that the progress in the field of materials (mainly nano-engineered materials like carbon nanotubes) will make space elevators possible in the future.

Terraforming An astro-engineering enterprise aimed at the transformation of the physical and environmental characteristics of the surface of a planet to make it suitable for supporting human life. The term *terraforming*, introduced by Isaac Asimov, has been widely used in science fiction, but now the possibility of terraforming planets, and initially Mars, is seriously considered. Terraforming a planet rises heated ethical arguments, in particular if the planet has any indigenous life that is likely to be destroyed in the process. One of the main points is whether it will ever be possible to be absolutely certain that no indigenous life exists on a planet.

Warp Drive Hypothetical faster than light (FTL) propulsion system. It is thought to be consistent with relativity, since FTL travel occurs not by achieving a speed greater than that of light, but by producing a distortion of the spacetime so that the distance traveled is actually shorter than that existing between the start and end point of the travel. It is much used by science fiction writers (e.g., in Star Trek movies) but is also studied by serious scientists.

References

1. G. Dyson, *Project Orion, The Atomic Spaceship 1957–1965*, (Penguin Books, London, 2002)
2. J. Dewar, *The Nuclear Rocket*, (Apogee Books, Burlington, 2009)
3. M.G. Millis, *Breakthrough Propulsion Physics Project: project Management Methods*, NASA/TM-2004-213406, (2004), E-14920
4. M. Alcubierre, "The warp drive: hyper-fast travel within general relativity." Classical Quant. Grav. **11**(5) (1994)

5. G. Genta, M. Rycroft, *Space, The Final Frontier?* (Cambridge University Press, Cambridge, 2003)
6. http://www.grc.nasa.gov/WWW/bpp/
7. R.M. Zubrin, D.A. Baker, *Mars Direct, A Proposal for the Rapid Exploration and Colonisation of the Red Planet, in Islands in the Sky*, (Wiley, New York, 1996)
8. http://science.nasa.gov/science-news/science-at-nasa/2000/ast07sep_1/
9. B. Finney, From sea to space, the Macmillan Brown lectures, Massey University, Hawaii Maritime Centre, (1992)
10. P.D. Ward, D. Brownlee, *Rare Earth*, (Copernicus, Springer, New York, 2000)
11. G. Genta, *Lonely Minds in the Universe*, (Copernicus, Springer, New York, 2007)
12. R. Penrose, *The Emperor's New Mind: Concerning Computers, Minds and the Laws of Physics*, (Oxford University Press, Oxford, 1989)
13. G. Genta, *Introduction to the Mechanics of Space Robots*, (Springer, New York, 2012)
14. R.A. Brooks, *Flesh and Machines*, (Pantheon Books, New York, 2002)
15. ABI Report, *Consumer and personal robotics*, http://www.Abiresearch.com
16. I. Yeoman, M. Mars, *robots, men and sex tourism*, Futures **44**(4), 365–371 (2012)
17. J. Carles, A. Dupleix, *Pierre Tehilard de Chardin*, (Centurion, Paris, 1993)
18. K.E. Drexler, *Engines of Creation, The Coming Era of Nanotechnology* (Oxford University Press, Oxford, 1990)
19. http://beyondturing.blogspot.it/2012/02/darwin-among-machines.html
20. F.J. Tipler, *The Physics of Immortality* (Macmillan, Basingstoke, 1994)
21. F. Valdes, R.A. Freitas Jr, *Comparison of reproducing and nonreproducing starprobe strategies for galactic exploration*, JBIS **33**, 402–406 (1980)
22. R.A. Freitas Jr., R.C. Merkle, *Kinematic Self-Replicating Machines*, Landes Bioscience, Georgetown, TX (2004)
23. S. Webb, *If the Universe is Teeming with Aliens… Where is Everybody?* (Copernicus, Springer, New York, 2002)
24. C. Sagan, W. Newman, *The solipsist approach to extraterrestrial intelligence*, Q. J. Roy. Astron. Soc. **24**(113), 115 (1983)

GPSR Compliance

The European Union's (EU) General Product Safety Regulation (GPSR) is a set of rules that requires consumer products to be safe and our obligations to ensure this.

If you have any concerns about our products, you can contact us on

ProductSafety@springernature.com

In case Publisher is established outside the EU, the EU authorized representative is:

Springer Nature Customer Service Center GmbH
Europaplatz 3
69115 Heidelberg, Germany